Dam Sector Protection and Homeland Security

ALSO IN THE CRITICAL INFRASTRUCTURE AND HOMELAND SECURITY SERIES

Energy Infrastructure Protection and Homeland Security
Food Supply Protection and Homeland Security
Transportation Protection and Homeland Security (forthcoming)
Government Facilities Infrastructure Protection and Homeland Security (forthcoming)

Dam Sector Protection and Homeland Security

FRANK R. SPELLMAN

Bernan Press
Lanham • Boulder • New York • London

Published in the United States of America
by Bernan Press, a wholly owned subsidiary of
The Rowman & Littlefield Publishing Group, Inc.
4501 Forbes Boulevard, Suite 200
Lanham, Maryland 20706

Bernan Press
800-462-6420
www.rowman.com

British Library Cataloguing in Publication Information Available

Library of Congress Cataloging-in-Publication Data

ISBN-13: 978-1-59888-873-7
eISBN-13: 978-1-59888-874-4

∞™ The paper used in this publication meets the minimum requirements of American National Standard for Information Sciences—Permanence of Paper for Printed Library Materials, ANSI/NISO Z39.48-1992.

Printed in the United States of America

For Loma Marie Spellman

Contents

Preface

As part of the well-received and highly acclaimed series on critical infrastructure and homeland security, *Dam Sector Protection and Homeland Security* is a reference source that is designed to serve and advise U.S. project designers, engineers, managers, superintendents, and supervisors of dams, navigation locks, levees, dikes, hurricane barriers, mine tailings and other industrial waste impoundments, and other similar water retention and water control facilities who want quick answers to complicated questions—to help employers and employees handle security threats they must be prepared to meet on a daily basis. In the post–September 11 world, the possibility of dam infrastructure terrorism—the malicious use of weapons and cyber intrusion to cause devastating damage to the dam infrastructure sector along with, literally, its cascading effects—is very real. Thus, the need is clear and so is the format and guidelines presented in this text to improve protection and resilience of the Dam Infrastructure Sector. This book describes the sector-wide process required to identify and prioritize assets, assess risk in the sector, implement protective programs and resilience strategies, and measure their effectives.

Within the context of this book, protective programs and resilience strategies cover a wide range of on-site and off-site efforts, such as implemented active or passive countermeasure and improving security protocols, hardening or retrofitting facilities to improve their performance, and implementing cybersecurity measures. The text points out that these efforts also include building redundancy and implementing backup systems to minimize disruptions or alleviate undesirable consequences, and incorporating hazard resistance into facility design.

It is important to point that our dam infrastructure (as is the case with the other sixteen critical infrastructures) cannot be made absolutely immune to all possible intrusions/attacks; thus, it takes a concerted, well-thought-out effort to incorporate security upgrades in the retrofitting of existing systems and careful security planning for all new energy infrastructure components. These upgrades or design features need to address issues of monitoring, response, critical infrastructure redundancy, and recovery to minimize risk to the facility/infrastructure. However, none of these approaches is or can be effective unless energy facility staff members at all levels of the chain of command are cognizant of the threats.

Dam Sector Protection and Homeland Security presents common-sense methodologies in a straightforward, almost blunt manner. Why so blunt? At this particular time, when dealing with the security of workers, family members, citizens, and society in general—actually, with our very way of life—politically correct presentations on security might be the norm, but in my view there is nothing normal or subtle about killing thousands of innocent people; mass murders certainly should not be expected; and the right to live in a free and safe environment is a reasonable demand.

This text is accessible to those who have no experience with the dam sector and/or homeland security. If you work through the text systematically, you will gain an understanding of the challenge of domestic preparedness—that is, an immediate need for a heightened state of awareness of the present threat facing the dam sector as a potential terrorist target. Moreover, you will gain knowledge of security principles and measures that can be implemented—adding a critical component not only to your professional knowledge but also give you the tools needed to combat terrorism in the homeland—our homeland, both by outsiders and insiders.

One final word to readers: this book is written in the conversational and engaging and reader-friendly style that is the author's trademark style.

Frank R. Spellman
Norfolk, Virginia

Prologue

Someone Left the Lid off of the Cookie Jar

My name is BM Williams II; yes, I am from the famous (some say, infamous) family of so-called in-house or homegrown terrorists who have tried to set this misguided country correctly back on its feet and to force it into a suitable direction for all of us freedom lovers. Yes, like my previous family members I have continued the trend toward fixing those things that are wrong with this country. In doing so—in accomplishing our ultimate goal—we believe in the Big Bang Effect . . . literally.

I sit here now incarcerated in a holding cell in King County, Seattle. I am able to tell you about my side of this sordid affair because I am in this holding cell awaiting the outcome of my fourth appeal . . . and I can also show you some photos along with telling you my side of the story because I have my evidence file here with me . . . in front of me now. . . . Yay! Some say I belong here, for sure. I have no doubt about my evilness and anti-establishment self. . . . I am who I am and say what I think and think what I say. I am not politically correct and proud of it. I am bad to the bone marrow . . . loving it. But I do not deserve to be locked up . . . but that is for further discussion and determination by the weirdoes that run things.

Okay, let's cut to the chase and I will explain to you grasshoppers out there in la-la land what I am locked up for and what my message to all of you is . . . for sure . . . part of my message is keep an eye on the cookie jar . . . what, you say? Patience, distorted, disturbed, and distressed ones . . . patience.

A few years ago I came to the conclusion that I needed to make a statement. The country is headed for bankruptcy and domination by foreign governments and we are being overrun by illegal immigrants many of whom would rather kill and rape us than acclimate themselves into our society. Yes, I needed to make a statement.

After some thought, my innate genius (all us Williams are born geniuses) came upon the idea of how to get the biggest bang for a medium-sized load of high explosives. My plan? Well, it certainly took some time and effort and lots of hiking energy to conceive the first one. The first one? Yes, but that will be explained when I get to the second one . . . are you grasshoppers with me so far? Ah, I thought so . . . let me continue. . . .

I decided to blow up a dam . . . a hydroelectric dam . . . a big one. I had visions of releasing millions of acre-feet of reservoir storage. This mammoth release, of course, was initially designed to wipe out, destroy, erase and kill everything downstream of the dammed reservoir. The concept was brilliant, thank you very much!

So, I had to decide what dam to take out. I decided it would be Hungry Horse Dam in Hungry Horse, Montana. Why, you ask? Hungry Horse Dam provided me with the isolated location that I thought was best to hide implementation of my intended destructive statement. Let me describe Hungry Horse Dam and its location . . . so that all you grasshoppers get the locational advantage and other pertinent facts or factoids . . . etc. . . . etc. . . . etc. . . . thank you very much . . . for sure. I have also included herein figures 1.1–1.4—I numbered them for simplicity.

The town of Hungry Horse, Montana, has a population of 900 and is a small community that is often overlooked (I liked this fact) by visitors as they make their way to Glacier National Park, which lies just ten miles or so to the east. I found that along with its isolation the best thing about Hungry Horse is the great huckleberry milk shakes that are offered by several roadside stands there. . . . I literally lived on those milkshakes while attempting to blow up the dam.

With regard to the dam, dimensionally it is 564 ft. (171.9 m.—I like the commie metric system, don't you?) tall; 2,115 ft. (736.4 m.) wide; 39 ft. (13.6 m.) thick at the top; and 330 ft. (114.8 m.) thick at base. It is located on the South Fork of the Flathead River and was designed to play an important role in providing water storage, flood control, hydroelectric power, and recreation

FIGURE 1.1
Hungry Horse Dam entrance area, Hungry Horse, Montana. Photo by F. R. Spellman.

FIGURE 1.2
Hungry Horse Dam face area, Hungry Horse, Montana. Photo by F. R. Spellman.

FIGURE 1.3
Hungry Horse Dam face area, power house, and South Fork of Flathead River, Hungry Horse, Montana. Photo by F. R. Spellman.

FIGURE 1.4
Hungry Horse Dam face, crest, with upper South Fork of Flathead River, Hungry Horse, Montana. Photo by F. R. Spellman.

for the area. Hungry Horse Dam contains almost 3.1 million cubic yards (2.4 million cubic meters) of concrete and is the eleventh highest concrete dam in the United States.

I do not think I need to describe the generating capacity or any of the so-called advantages the dam provides . . . let's face it; my plan was to destroy all that hyper bologna . . . so, again, let me continue with my brilliant plan to send Hungry Horse Dam to its ultimate appointment with crushing death and destruction.

In the weeks I spent checking out the Hungry Horse Dam surroundings and the dam itself, the one salient factor that grabbed my attention . . . surprised me, actually . . . but also generated a warm and fuzzy feeling throughout my very being was the dam's total lack of security.

Let me explain. I can hear your questions and expressions of doubt.

You might first ask: "Lack of security?"

"Yes, I say."

You go on: "In this age after 9/11?"

"Yes."

"Are you kidding us?"

"No."

"Explain, please."

"I shall—explaining is something that I am good at. . . ."

Like I stated earlier, I spent some time surveying my target, Hungry Horse Dam. This was no easy task. The reservoir itself totals 34 miles. I drove most of the 115 miles of reservoir road in close proximity to the dam—only 15 miles of which was pavement. But it was the visitor center that got my attention. On May 26 through May 29, 2015, I visited the visitor area and discovered to my total surprise that no one was around . . . not one site official or any security folks, that is. I did see a few visitors here and there but not that many. I noticed on the windowed side of the visitor center that all shades were pulled down tight and that there was a sign posted on the glass window. The sign stated that BLM (Bureau of Land Management) staff was in training and that the office and visitor center would be closed for the next few days. So no one was guarding the fort . . . so to speak. Moreover, I was shocked when I witnessed various van-sized vehicles and trucks cross the roadway top of the dam structure without anyone stopping to search them or to verify who they were.

Can you believe it?

Let me say it again: no security anywhere!

No security . . . heck, my kind of place to blow up, I thought.

My original plan was to disembark and launch (love those words, ha) the shoreline up-reservoir two miles in a large rubber raft with electric outboard motor (very quiet motor) and to pull three fully loaded larger rafts loaded with high explosives down to the dam face and then after placing them in place, I was to depart ashore and stage myself and the explosive trigger device at a safe distance from the dam face. . . . I am not a suicide type bomber . . . and I do not need 72 virgins, thank you very much. Anyway, that was my plan. All of it was to take place in darkness on one of those nights that was overcast with no full moon to light up the area . . . although I must say that I love full moons . . . don't you?

You see, my plan was to place the three fully loaded rafts right up alongside the crest or dam lip and hopefully blow a hole in 39 feet of cement . . . no easy task, for sure. And I realized this . . . so I changed my plan and added two more rafts designed to free flow into the first strike area where the first three rafts had exploded and hopefully provided enough explosive force to break the cement lip or weaken it to the point of making it easier for the second load of free floaters to do the follow-up damage.

But my plans changed.

Changed?

Yes.

Why?

Because I am no hero . . . no martyr . . . there is no cause worth dying for . . . like Patton said: Make the other SOBs die for the their country (or cause).

It was the glory hole that did it for me . . . no way Jose and Maria.

The dam was equipped with an emergency spillway for high water . . . in fact I found out later that it is the highest spillway or glory hole structure in the world (see figure 1.5).

Simply, the glory hole scared the heck out of me. You see I had this dream the night before I was to execute my brilliant plan. The river was very high . . . lots of snow melt from Glacier National Park. I launched my boat and the following rafts and headed down river toward the dam lip or crest. I had cut through the dam safety lines that were designed to protect the dam crest from floating logs and other debris . . . and I was headed right toward the crest

FIGURE 1.5
Hungry Horse Dam glory hole. Photo by F. R. Spellman.

when all of a sudden the sucking force on my craft exerted by the glory hole was pulling me right into the hole . . . heavens to murgatroid. . . . I was heading for the ultimate swallowing . . . chomped up by the sucking glory hole with the explosives on top of me . . . them going off and making my body the wall paper decoration of the interior remains of that glory hole . . . fortunately at the moment of being sucked into the glory hole I awoke . . . drenched in sweat and shaking like the proverbial leaf . . . you know, like those quaking aspen type leaves in the wind. . . . Needless to say, my plan to blow up Hungry Horse Dam was shelved . . . in a hurry. I am no hero . . . no martyr . . . I am a woman of style and substance and an American hero, of course. . . .

So, I moved on. But those haunting words of that genius (my personal hero) Bobby Burns were echoing and reverberating and resonating in my mindset . . . you know the ones: "The best laid schemes o' Mice an' Men, Gang aft agley. . . ." (more later . . . in plain English, of course).

As I stated, I moved on. I moved on to Seattle. Specifically, I moved on to the Ballard section of northwest Seattle . . . where I grew up. After realizing that blowing up Hungry Horse Dam was not an option for me anymore, I decided to move on to a more familiar target in Ballard; namely, I moved on to the Hiram M. Chittenden Locks . . . known locally as the Ballard Locks. The Locks sit at the west end of Salmon Bay, part of Seattle and Washington's Lake Washington Ship Canal. According to the U.S. Army Corps of Engineers and National Park Service, the locks basically serve three purposes:

- To maintain the water level of the fresh water Lake Washington and Lake Union at 20–22 feet (6.1–6.7 meters) above sea level (Puget Sound's mean low tide).
- To prevent salt water intrusion of sea water from Puget Sound with the fresh water of the lakes (mixing of sea water).
- To move boats from the water level of the lakes to the water level of Puget Sound, and vice versa.

In choosing the Locks as my next target, I also had three main purposes:

- To drain Lake Washington and Lake Union into Puget Sound to achieve true sea level . . . and to kill any man or fish that dared get in the way. . . .
- To mix salt and freshwater together . . . heck, I knew the surviving salmon and steelhead would not be offended.
- To destroy all them boats and occupants that transit the lakes and Sound.

My plan?

Well, grasshoppers, I could spend hours explaining to you what happened at the Ballard Locks . . . and explain in detail why I have been sentenced to life in prison without the possibility of parole . . . but an economy of words is called for here . . . so. . . .

It was all so simple, really . . . stroke of my genius, actually. Same as the Hungry Horse Dam plan.

I knew a place along the shore of Lake Union below the Aurora Bridge in the Fremont District . . . not far from my target, the Ballard Locks (of course).

On the chosen day of my attack, I had fellow compatriots load up five large rubber rafts with high explosives. I towed them with a flat bottom boat powered by one of those high horsepower outboard motors. I left the shore with my explosive load on that Sunday morning and cruised down Lake Union toward the locks at ever increasing speed. My plan was to follow a large luxury cabin cruiser that was in front of me inside the locks and then unlatch the explosive load about 50 feet from the first lock's open gate. After I sped away from the explosive load, I would trigger the load . . . as soon as I unleashed my boat from the tow and was able to make a radical U-turn and then speed away from the locks . . . in the resulting mayhem and confusion of death and destruction from the explosives I would be able to slip onto the shoreline and quietly and unobtrusively disappear from sight. I have provided a few pictures of the locks . . . my target . . . in figures 1.6–1.10.

Well, my best laid plans did not work . . . when I U-turned my boat it flipped over throwing me askew without the triggering device in hand to ignite the explosives and I was summarily captured and arrested by a bunch of wimpy Coast Guards people (anyone afraid to sail the deep ocean blue is a wimp, in my mind) . . . and then the police got involved and here I sit . . . a failure like the rest of the family.

FIGURE 1.6
Entrance to Hiram M. Chittenden Locks, Ballard, Washington. Photo by F. R. Spellman.

FIGURE 1.7
Gates at Hiram M. Chittenden Locks, Ballard, Washington. Photo by F. R. Spellman.

FIGURE 1.8
Boats entering Hiram M. Chittenden Locks from Lake Union, Ballard, Washington. Photo by F. R. Spellman

FIGURE 1.9
Hiram M. Chittenden Locks after lake water in out locks lowered to bay level, Ballard, Washington. Photo by F. R. Spellman.

FIGURE 1.10
Figure 1.10 Shows outer locks water level lowered to bay level with gates ready to open and let boats transit from freshwater to salt water environment. Ballard, Washington. Photo by F. R. Spellman.

And now I can only think of that haunting refrain professed my hero, Bobby Burns . . . (in understandable English, of course).

> The best laid schemes of Mice and Men
> oft go awry,
> And leave us but grief and pain,
> For promised Joy!
>
> —Robert Burns (1785)

Ah, Bobby . . . so true, laddie . . . so true.

One final word while I sit here and rot . . . and await my ultimate demise . . . my family has found that most the country's critical infrastructure is poorly guarded . . . only our own family stupidity has prevented us from success . . . and, grasshoppers, keep in mind that my family members need only be successful once to deliver our message . . . for we have found that someone has left the lid to the cookie jar off. The Williams family is a large one . . . and we do like cookies.

1

Introduction

When I say that terrorism is war against civilization, I may be met by the objection that terrorists are often idealists pursuing worthy ultimate aims—national or regional independence, and so forth, I do not accept this argument. I cannot agree that a terrorist can ever be an idealist, in that the objects sought can ever justly terrorism. The impact of terrorism, not merely on individual nations, but on humanity as a whole, is intrinsically evil, necessarily evil and wholly evil.

—Benjamin Netanyahu, *Fighting Terrorism*

NON-INTENTIONAL VS. INTENTIONAL DAM INFRASTRUCTURE DISASTERS

Knowing that humankind cannot predict, at the present time, exactly when natural disasters such as floods, volcanic eruptions, earthquakes, tsunamis, and other geologic processes are going to occur is a given; simply, it is reality. However, it is important to note that none of these natural occurrences or events can be classified as disastrous unless there is a vulnerable population residing within the event area.

Now it is arguable, by some people, as to what the exact definition of "population" is. For example, when Mount St. Helens had its catastrophic North Slope lateral eruption (May 18, 1980) that killed 57 people, destroyed

200 houses, 27 bridges, 15 miles (24 km) of railways, and 185 miles (298 km) of highway, could these humans and material things be considered as population or part of population? Or do we just count humans? Moreover, we can ask the same question; that is, is the more than 4,000,000,000 board feet (9,400,000 m^3) of timber that was destroyed or damaged during the eruption part of the population? Then consider downwind of the volcano where many agricultural crops, such as wheat, apples, potatoes, and alfalfa, were destroyed by thick ash accumulation. Are these human-planted crops part of a population? Then there are the estimated 1,500 elk and 5,000 deer that were killed and an estimated 12 million Chinook and Coho salmon fingerlings that died when their hatcheries were destroyed (Tilling et al. 1990; USGS 2015). And how about the billions of insects killed? Are they part of the population? Would you list their demise as disastrous? Additionally, consider that the level of Spirit Lake (at the foot of Mount St. Helens) was raised to its present position of 200 feet above original level, and the 24 megaton thermal energy explosion (1,600 times the size of the atomic bomb dropped on Hiroshima) rerouted parts of the North and South Toutle Rivers (USGS 2014). Are any of these human- or nature-made structures part of a population, as we define it?

Depends.

Depends on what?

It depends on whether the bridges, houses, fisheries, apple farms, or timber areas were occupied by humans who were killed. Understand that when a bridge is destroyed it is bad. On the other hand, if that same bridge is destroyed and it happens to take a van full of humans to their death, then we are talking disaster. At least, that is how the media will play it up and how affected relatives and loved ones will view it.

At this point in the presentation the reader is probably wondering what does the 1980 disastrous eruption of Mount St. Helens and other natural disasters have to do with dam infrastructure protection. Well, consider that when Mount St. Helens erupted on May 18, 1980, an estimated 40,000 young salmon were lost when they swam through turbine blades of hydroelectric generators when reservoir levels were lowered along the Lewis River to accommodate possible mudflow and flood waters. The hydroelectric dams near Mount St. Helens were not damaged but they were required to adjust flow and kill swarms of fish to prevent further downstream damage caused by the eruption (Tilling et al. 1990; USGS 2015).

The Mount St. Helens eruption and its subsequent impact on the local dam infrastructure is a recent event; however, by no means was it the first dam affected by a disaster. For example, the Johnstown Flood and the St. Francis Dam come to mind when one thinks about or studies dam disasters. Before discussing these two disasters (lots of people were killed) it is important to point out right up front that neither event was the result of natural causes; instead, they were the result of faulty human-made decisions, design, engineering, and management.

Johnstown Flood

On May 31, 1889, the Great Flood of 1889 (nationally, the Johnstown Flood) occurred as a result of the catastrophic failure of the South Fork Dam on the Little Conemaugh River 14 miles (23 km) upstream of the town of Johnstown, Pennsylvania. After several days of extremely heavy rainfall, the dam broke, unleashing 20 million tons of water (18 million m^3) from Lake Conemaugh. The flood killed 2,209 people and caused $425 million (in 2012 dollars) in damage (World Digital Library 2015; Perkins 2009; Gibson 2006).

The question is, of course, what caused the dam to break? The exact cause of the dam break has been an ongoing argument for decades. The courts eventually ruled the dam break to have been an Act of God. The survivors were granted no compensation. Again, notwithstanding God's influence, let's get back to the basic question: What caused the dam to break? Over the decades some have blamed the managers of the dam for the break. The managers supposedly modified the dam in ways to make the area more attractive to hunters and fishermen who belonged to a select club in the area—a club that was composed of members who were responsible for maintaining the dam in good condition. Two things are certain. First, this human-made dam broke and killed 2,209 people and damaged numerous properties. Second, if the dam had not been constructed in the first place, the killer flood could not have occurred as it did. An Act of God or human-made error would not have been applicable.

St. Francis Dam

The name "Mulholland" is familiar to many people. Maybe the name is familiar because of the song "Mulholland Drive" or the 2001 movie *Mulholland Drive*. Or perhaps the name is well known to those who have lived in or traveled through the Los Angeles area, the San Fernando Valley, and/or Los Angeles and Ventura Counties. The name is derived from pioneering LA civil engineer William Mulholland. What is not well known, especially to those outside the LA area, is William Mulholland's connection to the St. Francis Dam. The St. Francis Dam was built between 1924 and 1926 under the direction of the Bureau of Water Works and Supply managed by William Mulholland. The purpose of the dam was to store and regulate water for the City of Los Angeles; it was an integral part of the aqueduct water supply system. The dam itself was located about 10 miles north of what is present-day City of Santa Clara and about 40 miles from downtown Los Angeles.

Just before midnight on March 12, 1928, the dam catastrophically failed and the fast-moving floodwaters took the lives of an estimated 600 people (occupying second place for greatest loss of life in California after the 1906 San Francisco earthquake and fire) (Pollack 2010). The disaster marked the end of Mulholland's career (Mulholland 1995).

As you might imagine, the collapse of the St. Francis Dam and the resulting death and destruction that resulted was investigated by multiple investigative teams of engineers and geologists, committees and agencies and is still studied (as it should be) in college-level civil engineering classes today. The initial investigations were conducted as quickly as possible because the startup of construction of Hoover Dam was in the offing and there were many people concerned about building the much larger Hoover Dam; they did not want a similar catastrophe to occur.

With so many different groups investigating the dam catastrophe, it is to be expected that several different opinions and findings would be rendered. And that was the case, but in general the findings were pretty much in line with each other. The investigators reached consensus on the inadequacy of the dam foundation. One report stated that "the foundation under the entire dam left very much to desired . . . with such a formation [geologic], the ultimate failure of this dam was inevitable, unless water could have been kept from reaching the foundation" (Jackson and Hundley 2004). The committee appointed by the City Council of Los Angeles to investigate the cause of the failure of the St. Francis Dam reported its conclusions as follows:

1. The type and dimensions of the dam were amply sufficient if based on suitable foundation.

2. The concrete of which the dam was built was of ample strength to resist the stresses to which it would normally be subject.
3. The failure cannot be laid to movement of the earth's crust (not an Act of God).
4. The dam failed as a result of defective foundations.
5. This failure in no way reflects the stability of a well-designed gravity dam properly founded on suitable bedrock.

WHEN DISASTERS ARE WAITING TO HAPPEN

Natural disasters occur. Nonetheless, some people choose to ignore them or shrug off the possibility of an occurrence in their own backyards. Take, for example, those who choose to live in a known floodplain. The question is why? Answers: lived there all their lives, like it there, used to it and have boats ready to go, have flood insurance, or maybe they just think the taxpayers will bail them out each and every time (and we do). And then there are those who live in areas underlined by earthquake faults; just sitting or working or living there waiting for the "big one"—just waiting for disaster to occur.

The point here is that we all know that some disasters cannot be avoided, no matter what. We are human and our self-protective mechanisms are limited to body armor, staying away from bad places, staying away from drugs/alcohol abuse, and living in so-called safe zones (if there are any).

When natural disasters do occur, like the aforementioned eruption of Mount St. Helens, the Johnston Flood, and the St. Francis Dam collapse, can these events fall within the parameters detailed by Benjamin Netanyahu (in this chapter's epigraph)? That is, were the eruption of Mount St. Helens, the Johnston Flood, and the St. Francis Dam collapse intrinsically evil, necessarily evil, and wholly evil?

Depends.

Depends on what?

Depends on one's point of view. Depends on the victims' and survivors' points of view.

Again, common sense tells us that Mother Nature is in charge and if she wants to shake us up a bit here and there, she will. Whoa to those who think they can prevent this.

When any natural disaster occurs that results in damage to dam infrastructure, we have to classify such occurrences as a non-intentional event—a non-intentional event not caused intentionally by any human hand, that is. This book is not about natural disasters. It is about events best described by Benjamin Netanyahu: The intentional damage or destruction of dam infrastructure, including hydropower plants, navigation locks, levees, dikes, hurricane barriers, mine tailings and other industrial waste impoundments, and other water retention facilities is intrinsically evil, necessarily evil, and wholly evil.

REFERENCES AND RECOMMENDED READING

Gibson, C. 2006. "Our 10 Greatest Natural Disasters." *American Heritage* (August/September, 2006).

Jackson, D. C., and N. Hundley. 2004. "Privilege and Responsibility: Will Mulholland and the St. Francis Dam Disaster." *California History* (Fall 2004), pp. 8–47.

Perkins, S. 2009, November 21. "Johnstown Flood Matched Volume of Mississippi River." *Science News*, 176 (11).

Pollack, A. 2010. "President's Message." *Heritage Junction Dispatch*, March–April. Santa Clara, CA.

Mulholland, C. 1995. *William Mulholland and the St. Francis Dam: St. Francis Dam Disaster Revisited.* Santa Clara, CA: Historical Society of Southern California.

Tilling, R., L. Topinka, and D. A. Swanson. 1990. *Eruption of Mount St. Helens: Past, Present, and Future.* http://vulcna.wr.usgs.gov/VolcanoesMSH/Publications/MSHPPF/MSH_past_ (accessed May 5, 2015).

USGS. 2014. *Mount St. Helens—From the 1980 Eruption to 2000.* Fact Sheet 036–00. U.S. Geological Survey. http://pubs.usgs.gov/fs036-00/ (accessed May 5, 2015).

USGS. 2015. *The Climatic Eruption of May 18, 1980.* U.S. Geological Survey (Special Interest Publication; Public Domain). Washington, DC.

World Digital Library. 2015. *Johnstown, Pennsylvania, 1904.* Library of Congress. http://wdl.org/en/item/9571/ (accessed May 5, 2015).

2

Critical Infrastructure

> We can't accommodate terrorism. When someone uses the slaughter of innocent people to advance a so-called political cause, at that point the political cause becomes immoral and unjust and they should be eliminated from any serious discussion, any serious debate.
>
> —Rudolph Giuliani, September 11, 2002

WHAT IS CRITICAL INFRASTRUCTURE?

For the United States of America, 9/11 was a slap in the face, a punch in the gut (actually, the ultimate sucker punch), and a most serious wake-up call. The 9/11 wake-up call generated several reactions on our part—obviously, protecting ourselves from further attack became (and hopefully still is) priority number one. In light of this important need (i.e., the survival of our way of life), the Department of Homeland Security was created. According to former secretary Tom Ridge, "You may say Homeland Security is a Y2K problem that doesn't end January 1 of any given year" (Henry 2002). And, according to Barack Obama (2007), the Department of Homeland Security does "the work that ensures no other family members have to lose a loved one to a terrorist who turns a plane into a missile, a terrorist who straps a bomb around her waist and climbs aboard a bus, a terrorist who figures out how to set off a dirty bomb in one of our cities."

Among other things, the new emphasis on homeland security pointed to the need to protect and enhance the security of the nation's critical infrastructure. Critical infrastructure can be defined or listed in many ways. Generally, governments use the term to describe material assets, systems, and services that are essential for the functioning of an economy and society and maintaining public confidence. Destruction or compromise of any of these systems or services would have a debilitating impact on the area either directly, through interdependencies or from cascading effects. For the purpose of this text, critical infrastructure is defined as those assets of physical, key resources, and computer- or service-based systems that are essential to the minimum operations of and economy and government. Critical infrastructures (in the author's opinion) are the following:

- agriculture
- banking and finance
- chemical and hazardous materials
- defense industrial base
- emergency services
- energy
- organizations
- postal services and shipping
- public health
- strategies and assessments
- telecommunications
- transportation
- water and water treatment systems
- dams
- information technology
- national monuments and icons
- commercial assets
- nuclear power plants

Although we did not list cyberspace and all ancillaries (excluding, of course, the listing of information technology) involved in or with digital operations (e-technology), in this current era we can state without equivocation, doubt, ambiguity, or vagary that the digital connection is the glue that

DID YOU KNOW?

More than 85 percent of the critical infrastructure within the United States is owned and operated by the private sector.

holds all critical infrastructure together. This is the case, of course, because all separate infrastructures are interconnected in one way or another. This may surprise you to some degree, but think about it—we are not speaking about rocket science here; no, instead we are speaking about the present reality of e-technology. It would be hard to imagine that any of the listed infrastructure sectors could operate today without e-technology.

For example, consider Grand Coulee Dam, which is obviously a stellar member of the U.S. Dam Sector and is part of a sector where reliability is critical; it must never go off line unexpectedly (Stocks 2014). Like many other Dam Sector infrastructures, for many years Grand Coulee used SCADA for electronic control and security. The supervisory control and data acquisition (SCADA) system is still used extensively in many different applications but it is considered to be an aging system and therefore for better, more accurate, and more modern control and security applications, Grand Coulee Dam was recently retrofitted with a new power control system and is now operational using Real-Time Innovations (RTI), the real-time communications platform for the Internet Things company. "The Grand Coulee Dam control system delivers extreme availability, fault tolerance, performance, security and wide-area communications. When fully deployed, it must control 24 power turbines and 12 pumps. It comprises four major power plants, three high voltage transmission switchyards, and a fully-redundant control room" (Stocks 2014).

We could go down the list of critical infrastructures and easily point out where e-technology and the specific industry interface. However, since this is a discussion focusing on the dam sector, we will keep the focus on e-technology as it relates or interfaces with the dam sector. If not familiar with the dam sector—dam projects, hydropower plants, navigation locks, levees, dikes, hurricane barriers, mine tailing and other industrial waste impoundments, and other similar water retention and water control facilities—you might be surprised to learn that the entire modern hydropower and distribution grid is operated by e-technology from computer operation stations (manned by people, of course) with various digital proximity switches and other devices strategically positioned throughout the process to operate generators, pumps, valves, and switches; monitor critical parameters; and provide automatic emergency shutdown procedures.

DAM SECTOR INFRASTRUCTURE[1]

Dam Sector Protection and Homeland Security describes the study, design, and implementation of precautionary measures aimed to reduce the risk to our dam infrastructure from both homegrown (in-house, domestic) and/or foreign terrorism. But before we go any further in this text, let's define what the text is about; namely, what dams are.

Dams—Facilities that include some of all of the following components to perform their intended purposes safely: conventional dam or retaining structure (the structural sections that hold back the water), reservoir (the body of water impounded by the dam), spillways (the structure that passes normal and/or flood flows in a manner that protects the structural integrity of the dam), outlet works (the combination of structures and equipment required for safe operation and control of water released from a reservoir), powerhouse (the structure that houses turbines, generators, and associated control equipment for the production of hydroelectricity), and canal/aqueduct (a constructed channel, usually open, that conveys water by gravity or is used for navigation).

The U.S. Department of Homeland Security (2010) points out that dam infrastructure is critical to the economy of the twenty-first century. The sector continuously provides a wide range of economic, environmental, and social benefits, including hydroelectric power, river navigation, water supply, wildlife habitat, waste management, flood control, and recreation, just to name a few. Some examples of the benefits derived from dam sector assets are discussed next:

1. The sections in this chapter are based on the U.S. Department of Homeland Security's *National Infrastructure Protection Plan: Dam Sector*. Retrieved from www.dhs.gov/nipp (2015).

- **Water Storage and Irrigation**: Dams create reservoirs throughout the United States that supply water for many industrial, municipal, agricultural, and recreational uses. Ten percent of American cropland is irrigated by using water stored behind dams and thousands of jobs are tied to producing crops grown with irrigation water.
- **Electricity Generation**: The United States is one of the largest producers of hydropower in the world, second only Canada. Dams in the United States produce more than 270,000 gigawatt-hours, contributing 7 percent of the nation's electricity and representing 70 percent of the nation's renewable energy generation.
- **"Black Start" Capabilities**: There are 4,316 megawatts of "incremental" hydropower available at sites with existing hydroelectric facilities. Incremental hydropower is defined as capacity additions or improved efficiency at existing hydro projects. During the August 2003 blackout in the Northeast, hydropower projects in New York and several other states were able to quickly start generating electricity, leading the way to restoring power to millions of people.
- **Recreation**: Dams and other sector assets provide prime recreational facilities throughout the United States. In 2002, a total of 105.7 million recreation user-days and -nights were provided at hydropower projects licensed by the Federal Energy Regulatory Commission (FERC). In addition, about 400 million people annually visit a project of the U.S. Army Corps of Engineers (USACE) while about 90 million people visit a project of the Bureau of Reclamation.
- **Navigation**: Navigational projects constitute an essential component of the U.S. water system, which includes 236 lock chambers at 192 lock sites owned and/or operated by Army Corps of Engineers. The main benefit of the inland and intracoastal navigation system is the ability to efficiently move large volumes of bulk commodities long distances. If the cargo transported on the inland waterways each year had to be moved by another mode, it would take an additional 6.3 million rail cars or 25.2 million trucks to carry the load. The ability to move more cargo per shipment makes barge transport both fuel-efficient and environmentally advantageous.
- **Flood Risk Reduction**: Many dams and levees function as flood control projects, thereby reducing the potential human health and economic impacts of flooding. Reservoirs and levees built by USACE reportedly

prevented more than $19 billion in potential damages during the 1993 Midwest Flood. USACE lever systems currently provide a 6:1 return on flood damages prevented compared to initial costs and robust levee systems that provide a 24:1 return on investment. Levees and hurricane barriers reduce flood damage to rural communities as well as major metropolitan areas.

- **Sediment Control**: Some dams enhance environmental protection by controlling detrimental sedimentation.
- **Impoundment of Mine Tailings and industrial Waste Materials**: More than 1,500 mine tailings and industrial waste impoundments controlled by dams facilitate mining and processing of coal and other vital minerals while protecting the environment.

The potential risks in the event of asset failures, regardless of cause, within the Dam Sector are considerable and could result in significant destruction, including loss of life, massive property damage, and severe long-term consequences. Simply put, without a stable dam infrastructure sector, health and welfare are threatened and the U.S. economy cannot function effectively.

As mentioned, the dam infrastructure sector includes subsections or elements such as dam projects, hydropower plants, navigation locks, levees, dikes, hurricane barriers, mine tailing and other industrial waste impoundments, and other similar water retention and water control facilities. It is important to note that these subsections or elements of the dam sector encompass physical, human, cyber, and security elements.

DAM SECTOR ELEMENTS

As mentioned, the Dam Sector is comprised of physical, human, cyber, and security elements. The multiplicity and complexity of sector asset functions dictates the importance of these elements as described below.

Physical Elements

Dams in the United States range in height from a few feet to more than 700 feet, and in length from a few tens of feet to more than 1 mile. Normal reservoir storage (i.e., the volume of water stored in a reservoir for a majority of the time) behind U.S. dams can range from a few acre-feet to 30 million acre-feet. Dams in the U.S. Territories range from 30 feet to 274 feet high, with reservoir sizes that range from 100 acre-feet to 55,000 acre-feet. Dams are also located near or along the borders with Canada and Mexico.

In addition to the main impounding structure, one or several appurtenant structures may be associated with the multiple purposes of the facility. Examples of the types of structures that could be appurtenant to a dam include free overflow and gated spillway structures, outlet works, water supply features, penstocks, power generation features, navigation locks, embankments associated with the main structure (also known as saddle dams, dikes, and freeboard dams), and different types of control structures. Each dam type and each appurtenant structure has its own purpose and varying degrees of vulnerability. In general, each functional component may be associated with different protective configurations.

Levees (and flood risk reduction systems in general) may feature multiple components, including embankment sections, as well as floodwall sections, pumps and pumping stations, interior drainage works, flood damage reduction channels, closure structures, penetrations (e.g., utilities and pipes), and transitions (i.e., sections where the materials constituting the levee system change from, for example, a conventional floodwall section to an earthen section).

Human Element

The safe operation and security of the nation's critical infrastructure depend heavily on the human element. Dam Sector facilities can be either staffed or unstaffed. Many smaller or remote assets have minimal requirements for operation and are operated without personnel on-site for lengthy periods of time (e.g., a week). Some facilities are operated remotely from central operations centers and may or may not require on-site personnel on a daily basis. Some are staffed on a "business day schedule" (i.e., eight- or ten-hour schedules for five to seven days a week). Some are larger or more complex facilities require staff to be present on a 24/7 basis.

Because of the specialized nature of the work performed in the Dam Sector, the workforce for sector assets such as hydropower facilities or navigation locks has considerable stability compared to other industries, and each facility has a relatively small workforce.

With regard to hydropower dams, the number of personnel required to operate and maintain the asset can vary significantly. The need for staff depends greatly on the complexity of the dam and benefits derived from it. Larger hydropower dams may require adjustments in operations several times per hour to respond to generation needs and economics. They may require sev-

DID YOU KNOW?

Because many of the Dam Sector operations require small workforces, owners and managers are usually well acquainted with their workers and schedules. As a consequence, the potential for criminal infiltration is very limited: imposters, infiltrators, or inside threats are easier to recognize than is the case with other types of infrastructure. In addition, due to this smaller and more easily recognized workforce, the work environment is significantly easier to monitor, and unusual activity, such as surveillance, terrorist attacks, and other criminal activities, is much easier to distinguish and respond to in the early stages of a potential situation. However, this may not be necessarily applicable to other types of facilities, such as levees, for example, which are spatially distributed and require easy and open access for maintenance purposes.

eral operators for each shift; a crew of maintenance and monitoring personnel; specialized staff that respond to hourly, daily, or weekly adjustments to power demands; marketing specialists; transmission line crews; and security personnel and guards. Simple, less complex facilities may require only one person to operate the structure and perform monitoring and maintenance duties. For example, recreational dams and stock ponds have minimal staffing needs. Staffing at levees and flood control projects is somewhat similar to that of dams, but is more focused on operations and maintenance activities and will vary depending on water levels and weather conditions.

Because many of these positions are highly technical and specialized in nature they require personnel with three or more years of apprentice work or advanced training. Operations at some dams may also require the use of either full-time or corporate sharing of management and IT personnel. Personnel succession plans and continuity-of-operations plans help to ensure ongoing operation in the event of work stoppages and extended periods of illness, such as during pandemics. In addition, the schedule for noncritical operations can be modified to respond to situations that may be understaffed (e.g., establishing a deferred maintenance schedule).

Cyber Element

Efficiency and safety of dam sector operations and maintenance are constantly improving as technology in industrial control systems continue to advance. Control systems vary widely on the basis of age and generation of the system; thus, they also vary with respect to complexity and sophistication. Some systems are closed systems and use isolated networks, as well as proprietary communication protocols, while others use open architectures and common communication protocols and rely on the Internet. In addition, control systems at dams may also be connected to the electric power grid. Most dam owners, if not all, use various computer security methods for master terminal units, data servers, and historians, such as authentication procedures, encryption, firewalls, virus software, and anti-spyware.

Note that many of the industrial control systems used today were designed for operability and reliability during an era when security received low priority; that is, prior to 9/11 and the rise in worldwide terrorism. These systems operated in fairly isolated environments and typically relied on proprietary software, hardware, and/or communications technology. Infiltrating and compromising these systems often required specific knowledge of individual systems architecture and physical access to system components.

DID YOU KNOW?

As a general proposition, levee operations are conducted by smaller entities and are considerably less dependent on information technology than dams. They rely on basic telecommunications and information infrastructures that can be supplemented with manual operations.

In contrast, newer control systems are highly network based and use common and open standards for communication protocols, such as Internet Protocol addressable. Many controllers are Internet Protocol addressable. Asset owners and operators have gained immediate benefits by extending the connectivity of their systems and have increasingly adopted commercial off-the-shelf technologies that provide the higher levels of interoperability required in today's modern infrastructure. Standard operating systems such as Windows or UNIX are increasingly used in central supervisory stations, which are now typically connected to remote controllers via public-private networks provided by telecommunications companies. Common telecommunications technologies, such as the Internet, public-switched telephone networks, or cable or wireless networks, are often used.

Security Element

The unique characteristics of the Dam Sector influence the implementation of security measures. First, the spatial dimensions of dam projects are significantly larger than those of the facilities within other sectors. Second, there are often multiple components that are critical for project purposes that are not next door (non-adjacently located). Each one of these components may be associated with different security configurations. Third, many dams are located in remote areas; thus, a response force belonging to local law enforcement authorities may take longer to arrive once they receive an alarm. For some attack types, the appropriate response may not arrive in time to thwart the attempt. Finally, most dams have no armed guard forces.

The security configuration of sector assets may include a combination of features, such as fences and gates, vehicle-resistant perimeters, boat barriers, intrusion detection systems, electronic entry control systems, close-circuit television (CCTVs), appropriate lighting, patrols and guards, and hardening of select components. (Many of these security features are discussed later.)

In addition to these electronic and physical security features, Dam Sector partners also rely on robust security plants that define operations security procedures, such as information protection; threat- and suspicious-incident analysis centers; planned, called responses to varying threat levels; and coordination with law enforcement agencies. Owners and operators test Emergency Action Plans (EAPs) to help protect populations at risk and maintain response and recovery plans to bring project benefits back online rapidly.

THE BOTTOM LINE

Again, it is important to point out that the Dam Sector provides services that are essential to the U.S. economy and a major contributor to the so-called good life (standard of living) we presently enjoy. The Dam Sector is well aware of its vulnerabilities and is leading a significant voluntary effort to increase its planning and preparedness. Cooperation through industry groups has resulted in substantial information sharing of effective and best practices across the sector. Many sector owners and operators have extensive experience with infrastructure protection and have more recently focused their attention on cyber security. We hope that this book will aid in the prevention and mitigation of deliberate attacks, because we can't afford the economic consequences of a successful homegrown or foreign terrorist attack against Dam Sector facilities—or the threat to public health and safety and the environment.

REFERENCES AND RECOMMENDED READING

Carus, S. 2002. *Bioterrorism and Biocrimes: The Illicit Use of Biological Agents since 1900*. Washington, DC: Center of Counterproliferation Research, National Defense University.

CBO. 2004. *Homeland Security and the Private Sector*. Washington, DC: Congressional Budget Office.

FEMA, 2015. *Protecting Critical Infrastructure against Insider Threats*. http://emilms.fema.gov/IS0915/IABsummary.htm (accessed April 17, 2015).

FR. 2003. "Notice of Proposed Rulemaking." *Federal Register*, 68 (90).

Henry, K. 2002. "New Face of Security." *Gov. Security* (April), pp. 30–37.

Homeland Security. 2009. *National Infrastructure Protection Plan: Energy Sector*. www.dhs.gov/nipp (accessed April 24, 2009).

Horn, F. P. 1999. Statement Made before the United States Senate Emerging Threats and Capabilities Subcommittee of the Armed Services Committee. www.Senate.gov/~armed_servies/statement/1999/991027fh.pdf (accessed June 27, 2007).

Lane, J. 2002. Sworn Testimony, Congressional Field Hearing, House Committee on Government Reform, Abilene, KS.

Obama, B. 2007. *Homeland Security*. http://www.whitehouse.gov/agenda/homeland_security/.

Stocks, C. 2014. *Grand Coulee Dam*. Waterpowermagazine.com. www.andritz.com (accessed May 7, 2015).

U.S. Department of Homeland Security. 2010. *Dam Sector-Specific Plan: An Annex to the National Infrastructure Protection Plan*. Washington, DC: USDHS.

3

The Dam Sector

If inciting people to do that [9/11] is terrorism, and if killing those who kill our sons is terrorism, then let history be witness that we are terrorists.

—Osama Bin Laden, October 2001

When we speak of water, and its many manifestations we are speaking of that endless quintessential cycle that predates all other cycles. Water is our most precious natural resource; we can't survive without it. There is no more water today than there was yesterday; that is, no more this calendar year than 100 million years ago. The water present today is the same water used by all the animals that ever lived, by cave dwellers, Caesar, Cleopatra, Christ, da Vinci, John Snow, Teddy Roosevelt, and the rest of us—again there is not one drop more or one drop less of water than there has always been. This life-giving cycle, though it is a unique blend of thermal and mechanical aspects, is dependent on solar energy and gravity for its existence. Nothing on Earth is truly infinite in supply, but the energy available from water sources, in practical terms, comes closest to that ideal.

—F. R. Spellman (2008)

Given enough time, there is nothing, absolutely nothing, in this world that is safe from water.

—Frank R. Spellman (2014)

As mentioned earlier, the Dam Sector is composed of assets that include dam products, hydropower generation facilities, navigation locks, levees, dikes, hurricane barriers, mine tailings, other industries waste impoundments, and other similar water retention and water control facilities. The Dam Sector is a vital part of the nation's infrastructure and provides a wide range of economic, environmental, and social benefits, including hydroelectric power, river navigation, water supply, wildlife habitat, waste management, flood control, and recreation.

The Dam Sector has interdependencies with a wide range of other factors, including:

- **The Emergency Services Sector**—Emergency response sometimes rely on Dam Sector assets for firefighting water supply, emergency water supply, and waterway access during a significant disaster where access by land is impossible.
- **The Energy Sector**—Hydropower dams produce approximately 8 to 12 percent of the nation's power needs.
- **The Food and Agriculture Sector**—Dam Sector assets are a source of water for our nation's food and agriculture production.
- **The Transportation Systems Sector**—Dams and locks manage navigable waters throughout inland waterways for shipping and recreation.
- **The Water and Wastewater Systems Sector**—Water and wastewater management systems rely on Dam Sector assets to provide water to large populated areas and commercial facilities.

DAM COMPONENTS AND TYPES

Dams are facilities that include some or all of the following components to perform their intended purposes: a **conventional dam** or **impounding structure** (structural sections that hold back the water), a **reservoir** (a body of water impounded by the dam), **spillways** (structures that facilitate the discharge of normal and/or flood flows in a manner that protects the structural integrity of the project), **outlet works** (a combination of structures and equipment required for safe operation and control of water released from a reservoir), a **powerhouse** (a structure that houses turbines, generators, and associated control equipment for the production of hydroelectricity), **penstocks** (pipelines or conduits used to convey water under pressure to turbines or pump units), and **canals/aqueducts** (constructed channels, usually open, that convey water by gravity and also may be used for navigational purposes).

In general, canals convey water with an open channel from a source to a destination and serve a variety of functions. Some canals, constructed as part of water supply or transportation projects, have an embankment section forming one of the sides of the water conveyance structure and acting as a water retention structure. Water can be conveyed for a variety of uses (e.g., domestic, municipal, industrial, or agricultural water supply) or as a means of transportation for boats, barges, or other watercraft. In some cases, dams also include navigation locks.

Dams may be classified according to the type of construction materials used, the methods of construction, the slope or cross-section, the way that the structure resists the forces of the water pressure behind it, the means used for controlling seepage, the purpose, and the hazard potential. Materials used for construction include earth, rock, tailings from mining or milling, concrete, masonry, steel, timber, miscellaneous material such as plastic or rubber, or any combination of these materials. The two most common types of dams are embankment dams and concrete dams (ASDSO 2015).

Embankment dams are the most common type of dam in use today. An embankment dam is termed an "earthfill" or "rockfill" dam, depending on whether it is composed of compacted earth or mostly compacted or dumped rock. The ability of an embankment dam to resist the reservoir water pressure is primarily a result of the mass weight and the type and strength of the materials from which the dam is made.

Concrete dams may be categorized as gravity or arch dams. In concrete gravity dams, the most common form of concrete dam, the width of the concrete and the friction force along the foundation are used to resist the reservoir water pressure. Gravity dams are constructed of vertical blocks of concrete with flexible seals in the joints between the blocks. A buttress dam is a specific type of gravity dam in which the large mass of concrete is reduced, and the forces are diverted to the dam foundation through vertical or sloping buttresses. Concrete arch dams are characterized by thinner cross-sections in which the reservoir water pressure is carried laterally into the abutments. The shape of the arch may resemble a segment of a circle or an ellipse, and the arch may be curved in the vertical plane as well. Such dams are usually constructed of a series of thin vertical blocks that are keyed together; barriers to

Table 3.1. Ownership of U.S. Dams

Owner	*Percentage of Owners*
Private	65%
Local Government	20%
State Government	5%
Federal Government	4%
Unknown	4%
Public Utility	2%

stop water from flowing are provided between the blocks. Variations or arch dams include multi-arch dams in which more than one curved section is used and arch-gravity dams that combine some features of the two types of dams.

The National Inventory of Dams (NID) currently lists 82,642 dams in the United States; more than 70,000 of them are state-regulated, 4,639 are regulated by federal agencies, and the remainder are not regulated by any government agency. The ownership of dams in the United States makes them a unique part of the national infrastructure. Although most infrastructure facilities, such as roads, bridges, and sewer (wastewater) systems, are owned by public entities, the majority of dams (almost 65 percent) in the United States are owned by private entities. Table 3.1 illustrates the ownership structure and table 3.2 the primary purpose of dams in the United States.

Although table 3.2 shows that only about 2 percent of the dams within the United States are used for hydroelectric production purposes, it is important to point out that with regard to the importance of critical infrastructure applications the production of electrical power and hydropower (i.e., the harnessing of water to perform work) in general is just as important as flood control and water supply. The power of falling water has been used in industry for thousands of years over the ages as listed in table 3.3.

Table 3.2. Dams by Primary Purpose

Primary Purpose	*Percentage of Purpose*
Recreation	34%
Flood Control and Storm Water Management	17%
Fire Protection, Stock, or Small Farm Pond	15%
Irrigation	10%
Water Supply	7%
Other	7%
Unknown	4%
Hydroelectric	2%
Fish and Wildlife Pond	>2%
Tailings	>1.5%
Debris Control	>1.0%
Navigation	<0.5%

Table 3.3. History of Hydropower

Date	*Hydropower Event*
BCE	Hydropower used by the Greeks to turn water wheels for grinding wheat into flour, more than 2,000 years ago.
Mid-1770s	French hydraulic and military engineer Bernard Forest de Belidor wrote a four-volume work describing vertical- and horizontal-axis machines.
1775	U.S. Army Corps of Engineers founded, with establishment of Chief Engineer for the Continental Army.
1880	Michigan's Grand Rapids Electric Light and Power Company, generating electricity by dynamo belted to a water turbine at the Wolverine Chair Factory, lit up 16 brush-arc lamps.
1881	Niagara Falls city street lamps powered by hydropower.
1882	World's first hydroelectric power plant began operation on the Fox River in Appleton, Wisconsin.
1886	About 45 water-powered electric plants in the U.S. and Canada.
1887	San Bernardino, California, opens first hydroelectric plant in the west.
1889	Two hundred electric plants in the U.S. use water power for some or all generation.
1901	First Federal Water Power Act.
1902	Bureau of Reclamation established.
1907	Hydropower provided 15% of U.S. electrical generation.
1920	Hydropower provided 25% of U.S. electrical generation. Federal Power Act establishes Federal Power Commission authority to issue licenses for hydro development on public lands.
1933	Tennessee Valley Authority established.
1935	Federal Power Commission authority extended to all hydroelectric projects built by utilities engaged in interstate commerce.
1937	Bonneville Dam, first Federal dam, begins operation on the Columbia River, Bonneville Power Administration established.
1940	Hydropower provided 40% of electrical generation. Conventional capacity tripled in United States since 1920.
1980	Conventional capacity nearly tripled in United States since 1900.
2003	About 10% of U.S. electricity comes from hydropower. There is about 80,000 MW of conventional capacity and 18,000 MW of pumped storage.
2016	Hydroelectricity accounts for 7% of total energy production.

Source: EERE (2008) and USGS (2016).

DID YOU KNOW?

The potential energy in a specific slug or parcel of water is expressed in newton-meters (N · m). The newton is the standard unit of force, equivalent to a meter per second squared (1 m/s^2). This is the product of the mass of the slug in kilograms, the acceleration of gravity in meters per second squared (about 9.8 m/s^2), and the elevation of the parcel in meters (the vertical distance it falls as its energy is harnessed). The equivalent kinetic-energy unit is the joule (J), which is in effect equal to a watt-second (W · s).

Table 3.4. U.S. Energy Consumption by Energy Source, 2007 (Quadrillion Btu)

Energy Source	*2007*
Total	**101.605**
Renewable	6.830
Biomass (biofuels, waste, wood and wood-derived)	3.615
Biofuels	1.018
Waste	0.431
Wood derived Fuels	2.165
Geothermal	0.353
Hydroelectric Conventional	**2.463**
Solar/PV	0.080
Wind	0.319

Source: EIA 2007. *U.S. Energy consumption by Energy Source.* Accessed 6/12/09 @http://www.eia.doe.gov/enaf/alternate/page/renew_energy_conump/table 1.html.

Table 3.4 highlights solar energy's quadrillion Btu ranking in current renewable energy source use. As pointed out in the table, the energy consumption by energy source computations is from the year 2007. Thus, the 0.080 solar/PV quadrillion Btu figure is expected to steadily increase to a higher level and this should be reflected in the 2008–2015 figures when they are released. Table 3.5 lists the largest U.S. hydroelectric power stations.

DID YOU KNOW?

One of the problems encountered in a hydraulic system is storing the liquid. Unlike air, which is readily compressible and is capable of being stored in large quantities in relatively small containers, a liquid such as water cannot be compressed. Therefore, it is not possible to store a large amount of water in a small tank—62.4 lbs of water occupies a volume of one cubic foot, regardless of the pressure applied to it.

Table 3.5. Largest U.S. Hydroelectric Power Stations

Name	*Year of Completion*	*Total Capacity (Megawatts, MW)*
Grand Coulee	1942/1980	6,809
Bath County PSP	1945	3,003
Chief Joseph Dam	1958/73/79	2,620
Robert Moses Niagara PP	1961	2,515
John Day Dam	1949	2,160
Hoover Dam	1936/1961	2,080
The Dalles Dam	1981	2,038

Source: Bureau of Reclamation. (http://www.usbr.gov/pn/news/02new/gcrecrd.html) Retrieved 5/11/15.

1 ft³ H_2O = 62.4 lb

Both cubic feet and pounds are used to describe a volume of water. There is a defined relationship between these two methods of measurement. The specific weight of water is defined relative to a cubic foot. One cubic foot of water weighs 62.4 pounds. This relationship is true only at a temperature of 4°C and at a pressure of one atmosphere [known as standard temperature and pressure (STP)—14.7 lbs. per square inch at sea level containing 7.48 gallons]. The weight varies so little that, for practical purposes, this weight is used from a temperature 0°C to 100°C. One cubic inch of water weighs 0.0362 pounds. Water one foot deep will exert a pressure of 0.43 pounds per square inch on the bottom area (12 in × 0.0362 lb/in³). A column of water two feet high exerts 0.86 psi, one 10 feet high exerts 4.3 psi, and one 55 feet high exerts

$$55 \text{ ft} \times 0.43 \text{ psi/ft} = 23.65 \text{ psi}$$

A column of water 2.31 feet high will exert 1.0 psi. To produce a pressure of 50 psi requires a water column

$$50 \text{ psi} \times 2.31 \text{ ft/psi} = 115.5 \text{ ft}$$

Remember:

The important points being made here are:

1. 1 ft³ H_2O = 62.4 lb
2. A column of water 2.31 ft high will exert 1.0 psi

Another relationship is also important:

$$1 \text{ gallon } H_2O = 8.34 \text{ pounds}$$

FLOW/DISCHARGE RATE: WATER IN MOTION

The study of fluid flow is much more complicated than that of fluids at rest, but it is important to have an understanding of these principles because the water in used in hydropower applications is nearly always in motion (e.g., the water used to propel turbine blades).

Discharge (or flow) is the quantity of water passing a given point in a pipe or channel during a given period. Stated another way for open channels: The flow rate through an open channel is directly related to the velocity of the liquid and the cross-sectional area of the liquid in the channel.

$$Q = A \times V \tag{3.1}$$

where

Q = Flow—discharge in cubic feet per second (cfs)
A = Cross-sectional area of the pipe or channel (ft^2)
V = water velocity in feet per second (fps or ft/sec)

Example 3.1

Problem:

The channel is 6 feet wide and the water depth is 3 feet. The velocity in the channel is 4 feet per second. What is the discharge or flow rate in cubic feet per second?

Solution:

$$\text{Flow, cfs} = 6\text{ ft} \times 3\text{ ft} \times 4\text{ ft/second} = 72\text{ cfs}$$

Discharge or flow can be recorded as gallons/day (gpd), gallons/minute (gpm), or cubic feet (cfs). Flows treated by many hydropower systems are large, and often referred to in million gallons per day (MGD). The discharge or flow rate can be converted from cfs to other units such as gallons per minute (gpm) or million gallons per day (MGD) by using appropriate conversion factors.

RESERVOIR STORED ENERGY

One of the major components of a hydro-electric dam is the area behind the dam, its **reservoir**. The water temporarily stored there is called **gravitational potential energy**. The water is in a stored position above the rest of the dam facility so as to allow gravity to carry the water down to the turbines. Because this higher altitude is different than where the water would naturally be, the water is considered to be at an altered equilibrium. The result is stored energy of position; that is, gravitational potential energy. The water has the potential to do work because of the position it is in.

As shown in figure 3.1, gravity will force the water to fall to a lower position through the intake and the control gate. Installed within the dam, when the control gate is opened, the water from the reservoir goes through the intake and becomes translational kinetic energy as it falls through the next main part of the system, the penstock. Translational kinetic energy is the energy due to motion from one location to another. The water is moving (falling) from the reservoir toward the turbines through the penstock; the water is carried through the penstock's long shaft toward the turbines where the kinetic energy becomes mechanical energy. The force of the water is used to turn the turbines that turn the generator shaft. The generators convert the energy of water into electricity and then step-up transformers increase the voltage produced to higher voltage levels.

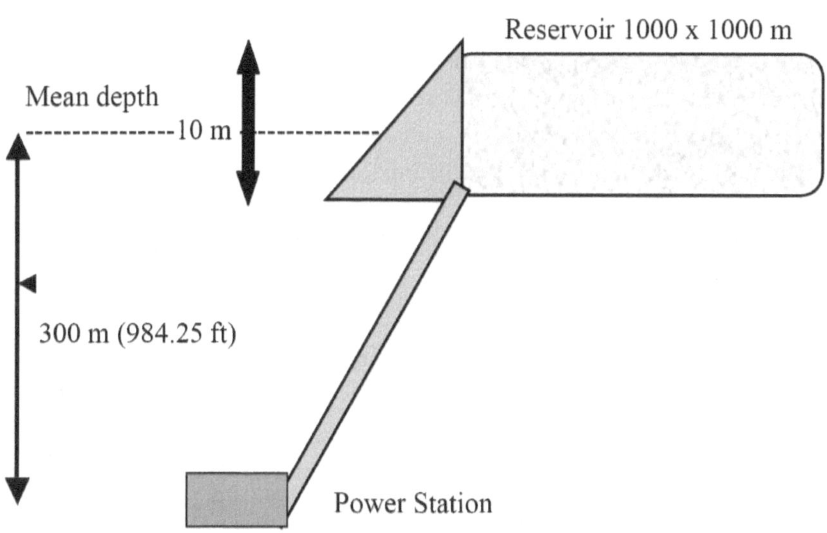

FIGURE 3.1
Schematic representation of a hydro scheme.
Adapted from N. K. Tovey. ENV-2E02 Energy Resources 2005. http://www2.env.uea.ac.uk.gmmc/energy (accessed May 10, 2015), pp. 108–109.

HYDRO TURBINES

According to EERE (2008), there are two main types of hydro turbines: Impulse and reaction. The type of hydropower turbine selected for a project is based on the height of standing water—as mentioned, this is referred to as "head"—and the flow, or volume of water, at the site. Cost, efficiency desired, and how deep the turbine must be set are other deciding factors.

Impulse Turbine—impulse turbines use the velocity of the water to move the runner and discharges to atmospheric pressure. The water stream hits each bucket on the runner. There is no suction on the down side of the turbine, and water flows out the bottom of the turbine housed after hitting the runner. An impulse turbine is generally suitable for high head, low flow applications.

Reaction Turbine—reaction turbines develop power form the combined action of pressure and moving water. The runner is placed directly in the water stream flowing over the blades rather than striking each individually. Reaction turbines are generally used for sites with lower head and higher flows than compared with impulse turbines.

NAVIGATION LOCKS

The term "navigation lock" is used to denote navigation projects that include the locks themselves (single or multiple chambers), as well as the associated dam (unless located in a canal), and a number of appurtenant structures such as approach walls, control rooms, and administration and maintenance buildings. The U.S. waterway system includes 236 locks chambers at 192 lock sites owned and/or operated by U.S. Army Corp of Engineers. Many navigation projects may serve multiple purposes in addition to the primary navigation function. For example, 46 lock-associated dams currently produce hydropower.

These facilities make many inland waterways viable, year-round transportation corridors. Natural river beds are often long, uneven downhill slopes with shallow areas and deep pools. Water depths can vary with seasonal rainfall and snow melt, and dry periods can significantly affect navigation. Locks on navigable rivers enable commercial and recreational river traffic to move safely from one pool level to the next. The lift of the locks—the difference between raising or lower from one pool to the next—can range from just a few feet to 110 feet (e.g., John Day Lock and Dam on the Columbia River, Oregon and Washington).

JOHN DAY DAM

The John Day Dam is a concrete gravity run-of-the-river dam spanning the Columbia River on the Washington/Oregon border (BPA, 2010). The dam features a navigation lock plus fish ladders on both sides and has the highest lift (110 feet) of any U.S. lock (USGS, 2015). Lake Umatilla is the reservoir impounded by the dam, and it runs 76.4 miles (123 kilometers) up the river channel to the foot of McNary Dam (USGS, 2015).

Multiple locks may be part of one navigable river system. The Tennessee River system is managed through a series of dams owned by the United States and operated by the TVA and the U.S. Army Corps of Engineers. The system's nine main and five auxiliary locks allow smooth movement on the river as it drops 513 feet in elevation from its beginning to the point at which it joins the Ohio River. The Mississippi River is made navigable through a total of 30 main and 4 auxiliary locks.

A principal value of the inland and intracoastal navigation systems is the ability to efficiently convey large volumes of bulk commodities moving long distances. Towboats push barges lashed together to form a "tow." A 15-barge tow is common on larger rivers with locks, such as the Ohio, upper Mississippi, Illinois, and Tennessee rivers. Such tows are an extremely efficient mode of transportation, moving about 22,500 tons of cargo as a single unit. A single 15-barge tow is the equivalent of about 225 rail cars or 870 tractor-trailer trucks. If the cargo transported on the inland water ways each year had to be moved by another mode, it would take an additional 6.3 million rail cars or 25.2 million trucks to carry the load.

The ability to move more cargo per shipment makes barge transport both fuel-efficient and environmentally advantageous. On average, a gallon of fuel allows 1 ton of cargo to be shipped 155 miles by truck, 413 by railway, and 576 miles by barge. As a result, cargo transported by barge results in an average transportation savings of about $11 per ton over the cost of shipping by alternative modes. Environmental advantage can also be measured by comparing the greenhouse gas (GHG) emissions produced during fuel consumption. On average, the tons of GHG emissions produced per million ton miles are 71.61 by truck, 26.88 by railway, and 19.27 by barge.

Coal is the largest commodity by volume moving on the inland waterways. The nation's electric industry depends on inland waterways for more than 20 percent of the coal that it consumes to produce electricity. Petroleum is the next largest group, including crude oil, gasoline, diesel fuel, heavy fuel oils, and asphalt. Another large group includes grain and other farm products, most of which move by waterway to ports on the Lower Mississippi or Columbia Rivers for export overseas. Sixty percent of the country's farm exports travel through inland waterways. Other major commodities include aggregates, such as stone, sand, and gravel used in construction; chemicals, including fertilizers; metal ores, minerals, and products, such as steel; and many other manufactured products.

LEVEES

The term "**levee**" denotes manmade flood risk reduction systems (e.g., embankment sections, floodwalls, closure structures, pumps, interior drainage works, and flood damage reduction channels) whose primary purpose is to furnish flood protection from seasonal high water and which are therefore subject to water loading for periods of only a few days or weeks a year. Embankments that are subject to water loading for prolonged periods (longer than normal flood protection requirements) or water loaded permanently should be designed in accordance with earth dam criteria rather than the levee criteria given in table 3.6.

Even though levees are similar to small earth dams they differ from earth dams in the following important respects: (a) a levee embankment may become saturated for only a short period of time beyond the limit of capillary saturation, (b) levee alignment is dictated primarily by flood protection requirements, which often results in construction on poor foundations, and (c) borrow is generally obtained from shallow pits or from channels excavated adjacent to the levee, which produce fill material that is often heterogeneous and far from ideal. Selection of the levee section is often based on the properties of the poorest material that must be used.

CAPILLARY ACTION

If we were to mention the term capillary action to the man or woman on the street, they might instantly nod their heads and respond that their bodies of full of them—that capillaries are the tiny blood vessels that connect the smallest arteries and the smallest of the veins. This would be true, of course. But in the context of water science, capillary is something different than capillary action in the human body.

Even if you've never heard of capillary action, it is still important in your life. Capillary action is important for moving water (and all of the things that are dissolved in it) around. It is defined as the movement of water within the spaces of a porous material due to the forces of adhesion, cohesion, and surface tension.

Surface tension is a measure of the strength of the water's surface film. The attraction between the water molecules creates a strong film, which among other common liquids is only surpassed by that of mercury. This surface tension permits water to hold up substances heavier and denser than it. A steel needle carefully placed on the surface of a glass of water will float. Some aquatic insects such as the water strider rely on surface tension to walk on water.

Capillary action occurs because water is sticky; thanks to the forces of cohesion (water molecules like to stay closely together) and adhesion (water molecules are attracted and stick to other substances). So, water tends to stick together, as in a drop, and it sticks to glass, cloth organic tissues, and soil. Dip a paper towel into a glass of water and the water will "climb" onto the paper towel. In fact, it will keep going up the towel until the pull of gravity is too much for it to overcome.

Table 3.6. Levees: Major and Minimum Requirements

Step	*Procedure*
1.	Conduct geological study based on a thorough review of available data including analysis of aerial photographs. Initiate preliminary subsurface explorations.
2.	Analyze preliminary exploration data and from the analysis establish preliminary soil profiles, borrow locations, and embankment sections.
3.	Initiate final exploration to provide a. Additional information of soil profiles. b. Undisturbed strengths of foundation materials. c. More detailed information on borrow areas and other required excavations.
4.	Using the information obtained in Step 3: a. Determine both embankment and foundation soil parameters and refine preliminary sections where needed, noting all possible problem areas. b. Compute rough quantities of suitable material and refine borrow area location
5.	Divide the entire levee into reaches of similar foundation conditions, embankment height, and fill material and assign a typical trial section to each reach.
6.	Analyze each trial section as needed for: a. Underseepage and through seepage. b. Slope stability. c. Settlement. d. Trafficability of the levee surface.
7.	Design special treatment to preclude any problems as determined from Step 6. Determine surfacing requirements for the levee based on its expected future use.
8.	Based on the results of Step 7, establish final sections for each reach.
9.	Compute final quantities needed, determine final borrow area locations.
10.	Design embankment slope protection.

With regard to levee design, numerous factors must be considered. These factors may vary from project to project, and no specific step-by-step procedure covering details of a particular project can be established. However, it is possible to present general, logical steps based on successful pact projects that can be followed in levee design and can be used as a base for developing more specific procedures for any particular project. Such a procedure is given in table 3.6.

The method of construction must also be considered. In the past levees have been built by methods of compaction varying from none to carefully controlled compaction. The local economic situation also affects the selection of a levee section. Traditionally, in areas of high property values, high land use, and good foundation conditions, levees have been built with relatively steep slopes using controlled compaction, while in areas of lower prop-

DID YOU KNOW?

Levees, as opposed to dams, are geospatially extended systems that can run for multiple miles; however, the levee system may consist of several segments under different governance units. This creates opportunities for multiple failure modes and points throughout the system. Proper levee maintenance is often overlooked because it appears to be fairly basic—repair erosion damage, protect against unwanted vegetation growth or animal burrowing, and so forth—so it may be given low priority or ignored completely.

erty values, poor foundations, or high rainfall during the construction season uncompacted or semicompacted levees with flatter slops are more typical. This is evident by comparing the steep slopes of levees along the industrialized Ohio River Valley with levees along the Lower Mississippi River which have much broader sections with gentler slopes. Levees built with smaller sections and steeper slopes generally require more comprehensive investigation and analysis than do levees with broad sections and flatter slopes whose design is more empirical. Where rainfall and foundation conditions permit, the trend in design of levees is toward sections with steeper slopes. Levee maintenance is another factor that often has considerable influence on the selection of a levee section.

Levee Types According to Location

Levees are broadly classified according to the area they protect as either urban or agricultural levees because of different requirements for each. As used in this book, urban and agricultural levees are defined as follows:

1. **Urban levees.** Levees that provide protection from flooding in communities, including their industrial, commercial, and residential facilities.

2. **Agricultural levees.** Levees that provide protection from flooding in lands used for agricultural purposes.

Levee Types According to Use

Some of the more common terms used for levees serving a specific purpose in connection with their overall purpose of flood protection are time in table 3.7.

Table 3.7. Classification of Levees According to Use

Type	*Definition*
Mainline and tributary levees	Levees that lie along a mainstream and its tributaries, respectively.
Ring levees	Levees that completely encircle (i.e., "ring") an area subject to inundation from all directions.
Setback levees	Levees that are built landward of existing levees, usually because the existing levees have suffered distress or are in some way being endangered, as by river migration.
Sublevees	Levees built for the purpose of underseepage control. Sublevees encircle areas behind the main levee which are subject, during high-water stages, to high uplift pressures and possibly the development of sand boils. They normally tie into the main levee, thus providing a basin that can be flooded during high-water stages, thereby counterbalancing excess head beneath the top stratum with the basin. Sublevees are rarely employed as the use of relief wells or seepage berms make them unnecessary except in emergencies.
Spur levees	Levees that project form the main levee and serve to protect the main levee from the erosive action of stream currents. Spur levees are not true levees but training dikes.

Causes of Levee Failures

The principal causes of levee failure are

1. Overtopping
2. Surface erosion
3. Internal erosion (piping)
4. Slides within the levee embankment of the foundation soils.

HURRICANE BARRIERS

A **hurricane barrier, flood barrier, surge barrier,** or **storm surge barrier** is a specific type of floodgate, designed to restrict the inflow of storm surges and extratropical storms into bays, estuaries, rivers and streams, and/or onto

HEAD

Head is defined as the vertical distance water must be lifted from the supply tank to the discharge, or as the height a column of water would rise due to the pressure at its base. A perfect vacuum plus atmospheric pressure of 14.7 psi would lift the water 34 feet. If the top of the sealed tube is opened to the atmosphere and the reservoir is enclosed, the pressure in the reservoir is increased; the water will rise in the tube. Because atmospheric pressure is essentially universal, we usually ignore the first 14.7 psi of actual pressure measurements, and measure only the difference between the water pressure and the atmospheric pressure; we call this *gauge pressure*. For example, water in an open reservoir is subjected to the 14.7 psi of atmospheric pressure, but subtracting this 14.7 psi leaves a gauge pressure of 0 psi. This shows that the water would rise 0 feet above the reservoir surface. If the gauge pressure in a water main were 120 psi, the water would rise in a tube connected to the main:

$$120 \text{ psi} \times 2.31 \text{ ft/psi} = 277 \text{ ft} \quad \text{(rounded)}$$

The *total head* includes the vertical distance the liquid must be lifted (static head), the loss to friction (friction head), and the energy required to maintain the desired velocity (velocity head).

$$\text{Total Head} = \text{Static Head} + \text{Friction Head} + \text{Velocity Head} \qquad (3.2)$$

the adjacent lowlands. These barriers consist of gates that are closed during storms from a fully open position or partially open position, depending on the degree of routine use that they also serve.

Hurricane barriers in the northeast that protect against flooding in large bays are normally left fully open and are closed only during a storm. Barriers in southern Florida serve a routine water control purpose and are completely closed during a hurricane event.

DID YOU KNOW?

Levee ownership and governance patterns can sometimes be politically, socially, and economically more complex than those for dams. Levee activities may be cost-shared among agencies at several levels of government, which can contribute to delays in repairs, and response and mitigation actions. Actual ownership, especially for older levees, may be unclear and may require lengthy litigation before the necessary repairs can be undertaken. Unclear ownership of levees and the levee segments that contribute to a levee system complicates the coordination and communication required for levee safety and resilience.

Hurricane barriers are typically large steel structures. Each of the movable gates that span the Providence River in Providence, Rhode Island, is 40 feet wide and weighs 53 tons. Hurricane barriers are usually composed of at least two gates—one on each bank of the channel—but more complex configurations may include multiple gates that can be independently raised or closed depending on the surge volume.

MINE TAILINGS AND OTHER INDUSTRIAL WASTE IMPOUNDMENTS

The mining industry, electric power generation, and manufacturing industrials often use impoundment structures to store and dewater waste. In the coal industry, embankment structures used for the disposal of the coal waste are referred to as slurry impoundments, in other extractive industries, the structures used for the disposal of fine mine waste are referred to as tailings ponds. In other industries, they may be referred to as surface or industrial waste impoundments. The primary goal of these impoundments is the storage of tailings or mine waste, which are transported suspended in water and settle out in the impoundment.

Tailings Dams[1]

In order to obtain the metals and other minerals needed for industrial processes, fertilizers, homes, cars, and other consumer products, large quantities of rock are mined, crushed, pulverized, and processed to recover metal and other mineral values. A fine grind is often necessary to release metals and minerals, so the mining industry produces enormous quantities of fine rock particles, in sizes ranging from sand-sized down to as low as a few microns. These fine-grained wastes are known as **tailings**.

Until recent decades, the majority of mines were small underground operations with correspondingly modest requirements for tailings disposal. Since that time, due to increasing demand, it has become economical to mine large lower-grade deposits by utilizing advances made by mining equipment manufacturers and developments in mining and mining technology. This has greatly increased the amount of tailings and other wastes generated by individual mining projects and by the mining industry as a whole (USEPA, 1994).

There are approximately 1,000 active metal mines in the United States (USGS, 2003). Many of these have a least one tailings impoundment and often several impoundments grouped together in cells. The U.S. Environmental Protection Agency (EPA) estimates that there may be several thousand tailings impoundments associated with active non-coal mining, and tens of thousands of inactive or abandoned impoundments.

By far the larger proportion of ore mined in most industry sectors ultimately becomes tailings that must be disposed of. In the gold industry, for example, only a few hundredths of an ounce of gold may be produced for every ton of dry tailings generated. Similarly, the copper industry and others typically mine relatively low-grade ores that contain less than a few percent of metal values; the residue becomes tailings. Thus, tailings disposal is a significant part of the overall mining and milling operation at most hardrock mining projects. There are several methods used for tailings disposal. These include disposal of dry or thickened tailings in impoundments or freestanding piles, backfilling underground mine working and open-pits, subaqueous disposal, and, the most common method, the disposal of tailings slurry in impoundments. Modern tailings impoundments are engineered structures for permanently disposing of the fine-grained waste from mining and milling operations. At some projects, tailings embankments reach several hundred feet in height and the impoundments cover several square miles.

1. The following section comes from USEPA (1994), *Design and Evaluation of Tailings Dams* (Washington, DC: U.S. Environmental Protection Agency).

Historically, tailings were disposed of where convenient and most cost-effective, often in flowing water or directly into drainages. As local concerns arose about sedimentation in downstream watercourses, water use, and other issues, mining operations began impounding tailings behind earthen dams, which themselves were often constructed of tailings and other waste materials. The impoundments served the dual purpose of containing the tailings and, particularly in the arid west, allowing the reuse of scarce water.

More recently, concerns have been raised about the stability and environmental performance of tailings dams and impoundments. Stability concerns are raised in part by the use of tailings material in tailings dams/embankments. To mitigate these concerns, mining operations often rely on a certain amount of controlled seepage to enhance stability in such embankments—which in turn affects environmental performance. Ritcey (1989) has speculated that the need for sound impoundments in the uranium industry "probably" accounts for much of the attention paid to impoundment design in other types of facilities. Perhaps triggered by the initial attention to uranium impoundments, the increasing concern for environmental performance has led to better engineering design of tailings dams in other mining industry sectors, for both stability and environmental performance. For instance, experience gained with leach pad liners is being transferred to linings for tailings ponds, and the use of synthetic lining materials is growing (although use of liners is still far from being the industry norm). In addition, the use of cyanide and other toxic reagents in mill processes has raised special concerns for some tailings and is leading to increased treatment prior to disposal as well as increased attention to containment. Finally, continuing concerns over acid mine drainage is resulting in a growing body of research and emerging concepts of long-term control or mitigation.

Inactive tailings impoundments also are receiving more attention due to the long-term effects of windblown dispersal, ground water contamination, and acid drainage. In many cases, the costs of remediation can be considerable, exceeding the costs of original design and operation of the tailings impoundment.

Design depends on the quantity and the individual characteristics of the tailings produced by the mining and milling operation, as well as the climatic, topographic, geologic, hydrogeologic, and geotechnical characteristics of the disposal site, and on regulatory requirements related to dam safety and to environmental performance. What may work for one type of tailings may not work for another type, and may not work for the same tailings at different sites. Hence each situation requires its own design process. The estimated quantity of tailings to be disposed of is particularly important given the evolving nature of most mining projects. Tailings quantity estimates are based on estimated reserves that change continuously as mine development professes. Accordingly, the final size and design of tailings impoundments may differ substantially from initial projections. This variance between the estimates and the realities presents major challenges to federal land managers and state permit writers, who are faced with reviewing and overseeing tailings impoundment planning, design, and performance, and to the general public, who may ultimately pay for miscalculations resulting in environmental damages.

Tailings Dams and Terrorism

Under normal everyday circumstances, calculation of the safety factor for a tailings embankment requires an analysis of the potential failure surfaces of the embankments. There are a number of common failure modes to which embankments may be vulnerable. These include slope failure from rotational slide, overtopping, foundation failure, erosion, piping, and liquefaction. Each failure mode may result in partial or complete embankment failure. Moreover, the commission of an act (or acts) of deliberate sabotage can not only lead to these detrimental results but at the very least can worsen or hasten damage effects. Simply, acts of terrorism have changed the calculus related to failure modes of tailings embankments and other Dam Sector infrastructure.

THE BOTTOM LINE

Given the many different assets within the Dam Sector, a number of failure modes and mission disruption scenarios could be theoretically within reach of terrorists, saboteurs, or other aggressors with the necessary capabilities. This potential is a matter of special concern in the case of high-consequence facilities and constitutes an important element in the sector risk profile. The combined effects of growing downstream development, insufficient maintenance funding, continuous aging, inadequate past design practices, and exposure to potential extreme events caused by natural hazards or deliberate aggressor actions drive a complex risk picture that could affect millions of people located in the path of sudden dam or levee failure.

REFERENCES AND RECOMMENDED READING

Association of State Dam Safety Officials (ASDSO). 2015. *Introduction to Dams.* www
.damsafety.org (accessed May 10, 2015).

Bender, M. D., G. E. Hauser, and M. S. Shiao. 1991. *Modeling Boone Reservoir to Evaluate Cost-Effectiveness of Point and Nonpoint Source Pollutant Controls.* Report No. WR28-1-31-107. Tennessee Valley Authority, Engineering Laboratory, Norris, TN.

BPA. 2010. *The Columbia River System Inside Story.* http://www.bpa.gov/power/pg/columbia_river_inside_sotry.pdf (accessed May 11, 2015).

Brookfield, H. C. 1989. *Sensitivity to Global Change: A New Task for Old/New Geographers.* Reading, United Kingdom: Norma Wilkinson Memorial Lecture, University of Reading.

Brundtland, G. H. 1987. *Our Common Future.* New York: Oxford University Press.

Commoner, B. 1971. *The Closing Circle: Nature, Man, and Technology.* New York: Knopf.

DOE. 2009. *Fossil Fuels.* U.S. Dept of Energy. http://www.energy.gov/energysources/fossilfuels.htm (accessed June 10, 2009).

EERE. 2008. *History of Hydropower.* www.eere.energy.gov/windandhydro/printable_versions/hydro_hisotry.html (accessed June 15, 2009).

Gibilisco, S. 2007. *Alternative Energy Demystified.* New York: McGraw-Hill.

Ginzburg, L. R. 1986. "The Theory of Population Dynamics: 1. Back to First Principles." *Journal of Theoretical Biology* 122, pp. 385–99.

Library of Congress. 2009. *The World's First Hydroelectric Power Plant.* http://www.americaslibrary.gov/eti-bin/page.cgi/jb/gilded/hydro (accessed June 15, 2009).

Lund, J. W. 2007. "Characteristics, Development and Utilization of Geothermal Resources." *Institute of Technology* 28 (2), pp. 1–9.

March, P. A., T. A. Brice, M. H. Mobley, and J. M. Cybularz. 1992. "Turbines for Solving the DO Dilemma." *Hydro Review* 11 (1).

Masters, G. M. 1991. *Introduction to Environmental Engineering and Science.* Englewood Cliffs, NJ: Prentice Hall.

Miller, G. T. 1988. *Environmental Science: An Introduction*. Belmont, CA: Wadsworth.

Reid, W., et al. 1988. *Bankrolling Successes*. Washington, DC: Environmental Policy Institute and National Wildlife Federation.

Spellman, F. R. 1996. *Stream Ecology and Self-Purification*. Lancaster, PA: Technomic Publishing Company.

Spellman, F. R. 2008. *The Science of Water*, 2nd ed. Boca Raton, FL: CRC Press.

Spellman, F. R. 2014. *The Science of Water*, 3rd ed. Boca Raton, FL: CRC Press.

Tovey, N. K. 2005. *ENV-2E02 Energy resources 2005 Lecture*. www2.env.ac.UK.gmmc/energy (accessed March 1, 2010).

UCS. 2008. *Land Use Changes and Biofuels*. Union of Concerned Scientists. http://www.ucsusa.org/clean_vehicles/technologies_and-feusl/biofuels/Land-Use-Changes-and-Biofules.ht...\ (accessed June 9, 2009).

UM. 2006. *Urbanization and Global Change*. University of Michigan. http://globalchange.umich.edu/globalchange2/current/lectures/urban_gc/ (accessed June 24, 2009).

USEPA. 1994. *Design and Evaluation of Tailings Dams*. Washington, DC: U.S. Environmental Protection Agency.

USDOE. 2008. *Advanced Hydropower Technology*. http://wwwl.eere.energy.gov/windlandhydro/printable_versions/hydro_advtech.html (accessed March 1, 2010).

USGS. 2015. Lake Umatilla. http://geonames.usgs.gov/apex/f?p=gnispq:3:::NO::{#_P3_FID:1513298 (accessed May 11, 2015).

USGS. 2016. Hydroelectric power water use. http//water.usgs.gov/edu/wuhy.html (last modified October 27, 2016).

U.S. Department of Homeland Security. 2010. *Dam Sector-Specific Plan: An Annex to the National Infrastructure Protection Plan*. Washington, DC: USDHS.

4

Dam Sector Security

> Terrorism has become the systematic weapon of a war that knows no borders or seldom has a face.
>
> —Jacques Chirac, 1986

> The only truly secure system is one that is powered off, cast in a block of concrete and sealed in a lead-lined room with armed guards.
>
> —Gene Spafford

While it is true that the U.S. Dam Sector has not been attacked on a large scale by terrorists, it is also true that the potential risk is more than conjecture. From several terrorism incidents on critical infrastructure in South America, the Middle East, and Africa, it is apparent that critical infrastructure is a preferred target of the terrorists. Moreover, Al-Qaeda has made it clear that it has an interest in attacking the American critical infrastructure system.

The obvious question might be: What makes the Dam Sector such an attractive terrorism target? Consider the following:

- Dam Sector components or assets are geospatially spread throughout the nation with little definition of boundaries.
- Much of the Dam Sector is equipment is one-of-a-kind large machinery which is difficult to replace in the short term.

- Several Dam Sector components such as navigation locks, substations, distribution power lines, and remote buildings were originally designed and constructed without concern for terrorist intrusion or destructive activities.
- Transmission towers and distribution components are open and widely dispersed.
- Many Dam Sectors systems are monitored and operated using under protected and aging SCADA systems.
- As with many of industries attempting to economize, many Dam Sector facilities assign responsibility for safety and security as a collateral duty to a line employee instead of employing a full-time certified safety and armed security professional.

ENERGY ASSETS AND SYSTEM PARAMETERS

The Department of Energy and Bureau of Land Management has identified six general asset or system characteristics that are important parameters for evaluating the vulnerabilities of Dam Sector Infrastructure and developing risk management programs (USDOE 2010).

- **Physical and location attributes.** Consideration of these assists the Dam Sector in developing consequence, vulnerability, and protective strategies.
- **Cyber attributes.** Like physical attributes, these assist the Dam Sector to evaluate consequences and vulnerabilities, and develop protective strategies. Cyber systems that link and help monitor and control the energy systems are increasingly recognized as a potential vulnerability.
- **Volumetric or throughput attributes.** These define the extent of any damage, depending on the utilized capacity of the systems, or points where the system may be capacity constrained.
- **Temporal/load profile attributes.** The Dam Sector has a strong temporal or time-dependent dimension affected by the season of the year and/or time of day.
- **Human attributes.** Highly trained and skilled personnel are key factors in a comprehensive Dam Sector security plan. The availability of skilled and experienced technical talent is a concern in the Dam Sector. Sustaining essential technical knowledge a critical to maintaining the sector's safety, reliability, and security.

Even though it was first considered an environmental issue, Dam Sector safety and security has been an issue of congressional interest for many years. Thus, it is important to ensure that the person designated as site safety/security professional be fully trained in the profession, experienced, and properly certified.

HOMELAND SECURITY DIRECTIVES

As a result of 9/11, the Homeland Security Department was formed. On matters pertaining to Homeland Security, Homeland Security Presidential Directives are issued by the U.S. president. Each directive has a specific meaning and purpose and is carried out by the U.S. Department of Homeland Security. Each directive is listed and summarized in table 4.1.

Table 4.1. Homeland Security Presidential Directives

HSPD–1: Organization and Operation of the Homeland Security Council. (White House) Ensures coordination of all homeland security-related activities among executive departments and agencies and promote the effective development and implementation of all homeland security policies.
HSPD–2: Combating Terrorism through Immigration Policies. (White House) Provides for the creation of task force which will work aggressively to prevent aliens who engage in or support terrorist activity from entering the United States and to detain, prosecute, or deport any such aliens who are within the United States.
HSPD–3: Homeland Security Advisory System. (White House) Establishes a comprehensive and effective means to disseminate information regarding the risk of terrorist acts to federal, state, and local authorities and to the American people.
HSPD–4: National Strategy to Combat Weapons of Mass Destruction. Applies new technologies, increased emphasis on intelligence collection and analysis, strengthens alliance relationships, and establishes new partnerships with former adversaries to counter this threat in all of its dimensions.
HSPD–5: Management of Domestic Incidents. (White House) Enhances the ability of the United States to manage domestic incidents by establishing a single, comprehensive national incident management system.
HSPD–6: Integration and Use of Screening information. (White House) Provides for the establishment of the Terrorist Threat Integration Center.
HSPD–7: Critical Infrastructure Identification, Prioritization, and Protection. (White House) Establishes a national policy for federal departments and agencies to identify and prioritize United States critical infrastructure and key resources and to protect them from terrorist attacks.
HSPD–8: National Preparedness. (White House) Identifies steps for improved coordination in response to incidents. This directive describes the way federal departments and agencies will prepare for such a response, including prevention activities during the early stages of a terrorism incident. This directive is a companion to HSPD-5.
HSPD–8 Annex 1: National Planning. Further enhances the preparedness of the United States by formally establishing a standard and comprehensive approach to national planning.

(continued)

Table 4.1. ***Continued***

HSPD–9: Defense of United States Agriculture and Food. (White House) Establishes a national policy to defend the agriculture and food system against terrorist attacks, major disasters, and other emergencies.
HSPD–10: Biodefense for the 21st Century. (White House) provides a comprehensive framework for our nation's Biodefense.
HSPD–11: Comprehensive Terrorist-Related Screening Procedures. (White House) Implements a coordinated and comprehensive approach to terrorist-related screening that supports homeland security, at home and abroad. This directive builds upon HSPD–6.
HSPD–12: Policy for a Common Identification Standard for Federal Employees and Contractors. (White House) Establishes a mandatory, government-wide standard for secure and reliable forms of identification issued by the Federal Government to its employees and contractors (including contractor employees).
HSPD–13: Maritime Security Policy. Establishes policy guidelines to enhance national and homeland security by protecting U.S. maritime interests.
HSPD–16: Aviation Strategy. Details a strategic vision or aviation security while recognizing ongoing efforts, and directs the production of a National Strategy for Aviation Security and supporting plans.
HSPD–18: Medical Countermeasures against Weapons of Mass Destruction. (White House) Establishes policy guidelines to draw upon the considerable potential of the scientific community in the public and private sectors to address medical countermeasure requirements relating to CBRN threats.
HSPD–19: Combating Terrorist Use of Explosives in the United States. (White House) Establishes a national policy, and calls for the development of a national strategy and implementation plan, on the prevention and detection of, protection against, and response to terrorist use of explosives in the United States.
HSPD–20: National Continuity Policy. (White House) Establishes a comprehensive national policy on the continuity of federal government structures and operations and a single National Continuity Coordinator responsible for coordinating the development and implementation of federal continuity policies.
HSPD–21: Public Health and Medical Preparedness. (White House) Establishes a national strategy that will enable a level of public health and medical preparedness sufficient to address a range of possible disasters.

Source: Department of Homeland Security (2008).

ASSESSING CONSEQUENCES

The potential physical and cyber consequences of any incident, including terror attacks and natural or human-made disasters, are the primary consideration in risk assessment. In the context of this book, consequence is measured as the range of loss or damage that can be expected.

The consequences that are considered for the national-level comparative risk assessment are based on the criteria as set forth in HSPD–7. These criteria can be divided into four main categories:

- **Human Impact**: Effect on human life and physical well-being (e.g., fatalities, injures).

- **Economic Impact**: Direct and indirect effects on the economy (e.g., costs resulting from disruption of products or services, costs to respond to and recover from the disruption, costs to rebuild the asset, and long-term costs due to environmental damage).
- **Impact on Public Confidence**: Effect on public morale and confidence in national economic and political institutions.
- **Impact on Government Capability**: Effect on the government's ability to maintain order, deliver minimum essential public services, ensure public health and safety, and carry out national security-related missions.

An assessment of all categories of consequence may be beyond the capabilities available for a given risk analysis. Most Dam Sector assets can be or are potentially associated with mass causalities, but may also have economic and long-term health and safety consequences.[1]

REPORTING SUSPICIOUS ACTIVITY—WHAT DAM SECTOR OWNERS AND OPERATORS CAN DO[2]

Suspicious incidents may actually constitute pre-operational (planning) activities and should be reported immediately. But in order to identify suspicious activity, you have to establish norms and be very familiar with all aspects of existing behaviors (personnel), procedures, and physical environment. Maintain awareness of your environment.

"Suspicious activities" are irregular or highly unusual behaviors that may be associated with pre-operations/preparatory surveillance, operational activities exploring or targeting a sector facility or system, or any possible violation of law or regulation that could compromise a sector facility or system, jeopardizing life or property. In detail, these activities may include:

- **Bomb Threat**—any threat of a bomb attack on property or *adjacent* properties.
- **Elicitation**—requests for information via telephone, e-mail, or even mail asking for: technical data about the asset or as security system, or information on hours worked and the number of personnel on duty during any given hour.

1. This section was from U.S. Department of Homeland Security (2006), *National Infrastructure Protection Plan* (Darby, PA: DIANE Publishing).
2. Adapted from the Department of Homeland Security website.

- **Overflight**—airborne crafts (small planes, helicopters, remotely controlled aircraft, and drones) that appear to be using a crossing pattern to survey the property or facility.
- **Surveillance**—personnel intending to gather information about the facility, its operations, or protective measures through photography, videotaping, sketching, or use of vision-enhancing devices (e.g., binoculars).
- **Test of Security**—testing of security systems, response and reaction times at a facility, which also includes any attempts to test or penetrate physical security barriers or procedures, and attempts to acquire information, uniforms, badges, passes, or other security-related documents.
- **Weapons Discovery**—discovery of weapons or a weapons cache on the property or adjacent properties

While this list of activities may seem obviously suspicious, consider the following examples that may at first seem to have a benign explanation:

- Showing unusual interest in a facility (e.g., observing through binoculars, taking notes, photography, drawing maps, or drawing structures of the facility).
- Persons parking, standing, or loitering in the same area on multiple days with no reasonable explanation for doing so.
- Persons not associated with the facility showing an increased interest in the surrounding area.

Any suspicious activity should be reported quickly with as much detail as possible. Report to the owner (if applicable), Homeland Security, closest FBI office, local law enforcement—or all of these based on location.

The Dam Sector Suspicious Activity Reporting (SAR) Tool

The Suspicious Activity Report (SAR) provides Dam Sector partners with the ability to report and retrieve information pertaining to suspicious activities that may be associated with pre-operation/preparatory surveillance, activities exploring or targeting a critical infrastructure facility or system, or any possible violation of law or regulation that could compromise a facility or system and could jeopardize life or property. This online tool provides an extra opportunity for real time situational awareness across the sector.

REFERENCES AND RECOMMENDED READING

FEMA. 2015. *Protecting Critical Infrastructure against Insider Threats.* http://emilms.fema.gov/IS0915/IABsummary.htm (accessed April 17, 2015).

Lees, F. 1996. *Loss Prevention in the Process Industries* 3:A5.1–A5.11. New York: Butterworth-Heinemann.

Sullivant, J. 2007. *Strategies for Protecting National Critical Infrastructure Assets: A Focus on Problem-Solving.* New York: Wiley & Sons.

U.S. Department of Energy. 2010. *Energy Sector-Specific Plan: An Annex to the National Infrastructure Protection Plan.* Washington, DC: USDOE.

U.S. Department of Homeland Security (DHS). 2015. *Dams Sector Suspicious Activity Reporting.* http://www.dhs.gov/dams-secotr-suspciious-activity-reporting (accessed May 14, 2015).

U.S. Department of Homeland Security (DHS). 2006. *National Infrastructure Protection Plan.* Darby, PA: DIANE Publishing.

5

Dam Sector Vulnerability Assessment

In God we trust, all others we monitor.

—Intercept Operator's motto

Vulnerability means different things to different people . . . many associate vulnerability with a specific set of human activities. . . .

— S. S. D. Foster

Terrorism is carried out purposefully, in a cold-blood, calculated fashion. The declared goals of the terrorist may change from place to place. He supposedly fights to remedy wrongs—social, religious, national, racial. But for all these problems his only solution is the demolition of the whole structure of society. No partial solution, not even the total redressing of the grievance he complains of, will satisfy him—until our social system is destroyed or delivered into his hands.

—Benjamin Netanyahu

One consequence of the events of 9/11 was Department of Homeland Security's (DHS) directive to establish a Critical Infrastructure Protection Task Force to ensure that activities to protect and secure vital infrastructure are comprehensive and carried out expeditiously. Another consequence is a heightened concern among citizens in the United States over the security of their energy infrastructure (i.e., the uninterrupted supply of electrical power and fuel to power vehicles and homes). As mentioned previously, the Dam Sector is classified as "vulnerable" in the sense that inherent weaknesses in its operating environment could be exploited to cause harm to the system. There is also the possibility of a cascading effect—a chain of events—due to a terrorist act affecting a Dam Sector provider, which causes corresponding damage (collateral damage) to other nearby users. In addition to significant damage to the nation's Dam Sector, entities using energy to produce finished products or provide critical life-saving services (e.g., medical centers and hospitals) that come under terrorist attack could result in loss of life; industry shutdown; catastrophic environmental damage to rivers, lakes, and wetlands; contamination of drinking water supplies; long-term public health impacts; undermined public confidence in government; and disruption to commerce, the economy, and our normal way of life.

VULNERABILITIES

For the purpose of this book and according to FEMA 2008, vulnerability is defined as any weakness that can be exploited by an aggressor to make an asset susceptible to hazard damage.

In addition, according to the Department of Homeland Security (DHS 2009), vulnerabilities are physical features or operational attributes that render an entity open to exploitation or susceptible to a given hazard. Vulnerabilities may be associated with physical (e.g., a broken fence), cyber (e.g., lack of a firewall), or human (e.g., untrained guards) factors.

The FBI (2009) reports that since 9/11 there have been a variety of threats suggesting that U.S. energy facilities are being targeted for terrorist attacks. Although the information often is fragmentary and offers little insight into the timing and mode of an attack, the October 6, 2002, operation against the French oil tanker *Limburg* in the Gulf of Aden (a small boat carrying explosives caused a rupture spilling 90,000 barrels out of the cargo of 400,000) suggests that Al-Qaeda is serious about targeting the energy sector and other critical infrastructures. Al-Qaeda appears to believe that an attack on oil and gas structures could do great damage to the U.S. economy. Moreover, terrorist plotters probably perceive infrastructure such as dams and power lines as having softer defenses than other facilities.

INSIDER THREAT VULNERABILITY

Security and risk managers (and all employees in general) working in or with the critical infrastructure sectors must be aware of the potential for insider threat vulnerability. The key word is AWARENESS. Awareness means that personnel within the critical infrastructure sectors must know how to identify and take action against insider threats. To achieve this critical goal, safety and security personnel must be provided with an overview of and be cognizant of common characteristics and indicators associated with malicious insiders and effective measures to counter insider threats.

Protecting against Insider Threats[1]

When analyzing threats to our nation's critical infrastructure, we tend to focus on malicious actions from outside actors. Of equal concern (and even more so in the author's view) are threats from an insider—someone who has been given legitimate access to information, systems, and resources. The measures we take to detect and protect against *external* threats may not be sufficient to address *internal* threats.

A malicious insider has access and inside knowledge of the organization and uses that knowledge with the intent to cause harm. The insider may be a current employee, a former employee, a service provider, an embedded hostile agent, or even a planted person inside who has been brainwashed and turned into a terrorist waiting for the moment of maximum impact to people and property.

> The insider threat to critical infrastructure is one or more individuals with the access or inside knowledge of a company, organization, or enterprise that would allow them to exploit the vulnerabilities of the entity's security, systems, services, products, or facilities with the intent to cause harm.
>
> —The President's National Infrastructure Advisory Council (2008)

1. Adapted from the FEMA website.

Given the importance of our nation's critical infrastructure, the actions taken by a malicious employee or service provider could have devastating consequences. Look at some actual examples of malicious insiders:

- In June 2015, at a U.S.-owned gas factory in Lyons, France, an employee citing allegiance to ISIS beheaded his supervisor and rammed a truck into the facility to cause an explosion.
- A service provider employee at a nuclear facility stole two 5-gallon containers of low-enriched uranium dioxide and then attempted to extort $100,000 by threatening to disperse the material in an unnamed U.S. city.
- A power company field engineer, angry with his supervisor, disabled protection systems at a substation and forced the shutdown of the entire network. More than 800,000 customers lost power as a result.
- Two municipal employees used their access credentials to sabotage the system controlling the traffic lights of a major city, causing widespread traffic delays. The damage took four days to repair.

Insider threats can be accomplished through either physical or cyber means and may involve any of the following:

- **Physical or information-technology sabotage**—Involves modification or damage to an organization's facilities, property, assets, inventory, or systems with the purpose of harming or threatening harm to an individual, the organization, or the organization's operations.
- **Theft of intellectual property**—Involves removal or transfer of an organization's intellectual property outside the organization through physical or electronic means (also known as economic espionage).
- **Theft of economic fraud**—Involves acquisition of an organization's financial or other assets through theft or fraud.
- **National security espionage**—Involves obtaining information or assets with a potential impact on national security through clandestine activities.

Insider Threats to Energy Infrastructure Sector

Insider threats to critical infrastructure can occur over many years and have devastating consequences. Malicious incidents have occurred through-

DID YOU KNOW?

The FBI testified in June 2012 that in the preceding year, economic espionage losses to the American economy totaled more than $13 billion. In the previous four years, the number of arrests the FBI made had doubled; indictments increased by a factor of five; and convictions increased by a factor of eight (FBI, 2012). In another survey, security professionals found that 43.2 percent of respondents attributed some loss at their organization to insiders. 46 percent of responds said the damage cause by insider attacks was more damaging than outside attacks (CSI, 2011).

out the energy infrastructure sector despite the severe provisions of the Economic Espionage Act of 1996: penalties can include fines of up to $500,000 per offense and imprisonment of up to 15 years for individuals, as well as forfeiture of any criminal proceeds. The fact is no fine or other penalty is going to deter a dedicated insider terrorist from performing his or her evil act(s).

One of the most insidious, ongoing and unnoticed examples, for months or even years, of an increasing insider threat is intellectual property theft. Intellectual property theft can seriously compromise our ability to protect ourselves from outside attack, as shown in the following examples:

- Greg Chung was a Boeing engineer who, from 1979 to 2006, stole trade secrets about the space shuttle, the Delta IV rocket, and the C-17 military cargo jet. He sent that sensitive information to China, his native country. He was sentenced in February 2010 to more than 15 years in prison.
- The supervisor of a maintenance crew with access to every office at a commercial research and development facility stole confidential research information and tried to sell it to a rival company. He pled guilty in 1997 to theft of trade secrets and was sentenced to 15 months in jail.

Common Characteristics and Traits of Malicious Insiders

Based on research conducted by the Software Engineering Institute at Carnegie Mellon University and the U.S. Secret Service National Threat Assessment Center, malicious insiders often are perceived or known to be difficult or high-maintenance employees who are:

- Obviously unhappy or extremely resentful.
- Having financial, performance, or behavioral problems.
- At risk (or perceived to be) for layoff or termination.

Keep in mind that not all malicious insiders fit this characterization. Insiders involved in national security espionage, for example, may exhibit few outward signs. In the majority of cases, however, management and/or human resources personnel were well aware of the employees and their issues prior to an incident.

The following motives and personal situations frequently are linked with malicious insiders:

- **Personal or Behavioral Problems**
  - Vulnerable to blackmail
  - Experiencing family or financial problems
  - Prone to compulsive or destructive behavior
  - Subject to ego or self-image issues
- **Personal Desires**
  - Seeking adventure or thrill
  - Seeking approval and returned favors
  - Professing allegiance
- **Workplace Issues**
  - Experiencing problems at work
  - Feeling anger or need for revenge

The following organizational factors have been known to encourage or present opportunities to potential malicious insiders:

- **Access and Availability**
  - Ease of access to materials and information
  - Ability to exit the facility or network with materials or information
- **Policies and Procedures**
  - Undefined or inadequate policies and procedures
  - Inadequate training
  - Lack of training
- **Time Pressure and Consequences**
  - Rushed employees
  - Perception of lack of consequences

Insider Activities and Behavior You May See

Insider threats may be detected through particular activities and behavior on the part of the insider. These activities and behaviors often will appear unusual or suspicious. Keep in mind there may be several explanations for a particular activity or behavior identified here, but when combined with other factors, certain activity or behaviors point toward a possible insider threat. A combination or confluence of indicators should not be ignored.

Unusual or suspicious insider activities and behavior can be described using the following categories:

- Inappropriate Interest or Acquisition
- Unauthorized or Unusual Computer Use
- Unusual Hours, Contacts, or Travel
- Secretive or Peculiar Behavior
- Personal or Financial Issues

So what can employers do? There are a range of preventative actions that can be taken:

- Clearly communicating and consistently enforcing security policies and controls.
- Ensuring that proprietary information and materials are adequately, if not robustly, protected.
- Routinely monitoring computer networks for suspicious activity.

- Ensuring security (including computer network security) personnel have the tools they need.
- Consulting with legal and law enforcement experts as needed to ensure compliance with the law.
- Becoming familiar with behavior and activities associated with malicious insiders.
- Documenting and evaluating incidents of suspicious or disruptive behavior.

Critical infrastructure organizations today employ a number of security measures to reduce the risk of insider threats—to stop the problem before it enters the door. For example, in the hiring process, using appropriate screening processes to select new employees. Employers can enlist employees as part of the solution, by educating them about security or other protocols and by encouraging and providing non-threatening, convenient ways for them to report suspicious in a confidential manner.

VULNERABILITY ASSESSMENT[2]

A **vulnerability assessment (VA)** involves an in-depth analysis of a facility's functions, systems, and site characteristics to identify weaknesses and lack of redundancy and to determine mitigations or corrective actions that can be designed and implemented to reduce the vulnerabilities. A vulnerability assessment can be a standalone process or part of a full risk assessment. During this assessment, the analysis of site assets is based on: (a) the identified threat; (b) the criticality of the assets; and (c) the level of protection chosen (i.e., based on willingness or unwillingness to accept risk).

Dam Sector members should routinely perform vulnerability assessments to better understand threats and vulnerabilities, determine acceptable levels of risk, and stimulate action to mitigate identified vulnerabilities. The direct benefits of performing a vulnerability assessment include:

- **Build and broaden awareness**—The assessment process directs senior management's attention to security. Security issues, risks, vulnerabilities, mitigation options, and best practices are brought to the surface. Aware-

2. Much of the information in this section is adapted from U.S. Department of Energy (DOE). 2002. *Vulnerability Assessment Methodology: Electric Power Infrastructure*. Washington, DC.

ness is one of the least expensive and most effective methods for improving overall security.

- **Establish or evaluate against a baseline**—If a baseline has been previously established, an assessment is an opportunity for a checkup to gauge the improvement or deterioration of security. If no previous baseline has been performed (or the work was not uniform or comprehensive), an assessment is an opportunity to integrate and unify previous efforts, define common metrics, and establish a definitive baseline. The baseline also can be compared against best practices to provide perspective on a facility's security posture.
- **Identify vulnerabilities and develop responses**—Generating lists of vulnerabilities and potential responses is usually a core activity and outcome of an assessment. Sometimes, due to budget, time, complexity, and risk considerations, the response selected for many of the vulnerabilities may be non-action, but after completing the assessment process these decisions will be conscious ones, with a documented decision process and item-by-item rationale available for revisiting issues at scheduled intervals. This information can help drive or motivate the development of a risk management process.
- **Categorize key assets and drive the risk management process**—An assessment can be a vehicle for reaching corporate-wide consensus on a hierarchy of key asset. This ranking, combined with threat, vulnerability, and risk analysis, is at the heart of any risk management process. For many, "Y2K" was the first time a facility- or organization-wide inventory and ranking of key assets was attempted. An assessment allows any organization to revisit that list form a broader and more comprehensive perspective.
- **Develop and build internal skills and expertise**—A security assessment, when not implemented in an "audit" mode, can serve as an excellent opportunity to build security skills and expertise within a group or organization. A well-structure assessment helps managers and personnel come together and share issues, experiences, and expertise. External assessors can be instructed to emphasize "teaching and collaborating" rather than "evaluating" (the traditional role). A long-term goal should be to move toward a capability for self-assessment.
- **Promote action**—Although various security efforts may already be underway, an assessment can crystallize and focus attention and resources on

solving specific and systemic security problems. Often the people "in the trenches" are well aware of security issues (and even potential solutions) but are unable to convert their awareness to action. An assessment provides an outlet for their concerns and the potential to bring these issues to the appropriate supervisory level and achieve action.

- **Kick off an ongoing security effort**—An assessment can be used as a catalyst to involve people throughout a facility or organization in security issues, establish permanent trouble-shooting teams, and harness the momentum generated by the assessment to build an ongoing institutional security effort.

The Vulnerability Assessment Process

For our purposes, the common elements of a vulnerability assessment for the Dam Sector might be as follows:

- Characterization of the Dam Sector, including its mission and objectives.
- Identification and prioritization of adverse consequences to avoid.
- Determination of critical assets (dam components, etc.) that might be subject to malevolent acts that could result in undesired consequences.
- Assessment of the likelihood (qualitative probability) of such malevolent acts from adversaries.
- Evaluation of existing countermeasures.
- Analysis of current risk and development of a prioritized plan for risk reduction.

Table 5.1 provides an overview of the elements included in the assessment process. The elements included in this overview are based on actual in-field experience and lessons learned.

In table 5.1, step 3 deals with identification of asset criticality. This is an important first step in any vulnerability assessment. Identifying asset criticality serves several functions:

- It enables more careful consideration of factors that affect risk, including threats, vulnerabilities, and consequences of loss or compromise of the asset.
- It enables more focused and through consideration of loss or compromise of the asset.

Table 5.1. Basic Elements in Vulnerability Assessments

Element	*Points to Consider*
1. Characterization of the energy entity, including its mission and objectives.	• What are the important missions of the system to be assessed? Define the highest priority services provided by the utility. Identify the industry's customers: ** General public ** Government ** Military ** Industrial ** Critical care ** Retail operations ** Firefighting • What are the most important facilities, processes, and assets of the system for achieving the mission objectives and avoiding undesired consequences? Describe the: ** Industry facilities ** Operating procedures ** Management practices that are necessary to achieve the mission objectives ** How the industry operates ** Treatment processes ** Storage methods and capacity ** Energy use and storage ** Distribution system In assessing those assets that are critical, consider critical customers, dependence on other infrastructures (e.g., chemical, transportation, communications), contractual obligations, single points of failure, chemical hazards and other aspects of the industry's operations, or availability of industry utilities that may increase or decrease the criticality of specific facilities, processes and assets.
2. Identification and prioritization of adverse consequences to avoid.	• Take into account the impacts that could substantially disrupt the ability of the system to provide a safe and reliable supply of energy. Dam Sector systems should use the vulnerability assessment process to determine how to reduce risk associated with the consequences of significant concern. • Ranges of consequences or impacts for each of these events should be identified and defined. Factors to be considered in assessing the consequences may include: ** Magnitude of service disruption ** Economic impact (such as replacement and installation costs for damaged critical assets or loss of revenue due to service outage) ** Number of illnesses or deaths resulting from an event ** Impact on public confidence in the energy supply ** Chronic problems arising from specific events ** Other indicators of the impact of each event as determined by the Dam Sector.

(*continued*)

Table 5.1. *Continued*

Element	*Points to Consider*
	Risk reduction recommendations at the conclusion of the vulnerability assessment strive to prevent or reduce each of these consequences.
3. Determination of critical assets that might be subject to malevolent acts that could result in undesired consequences.	• What are the malevolent acts that could reasonably cause undesired? ** Electronic, computer or other automated systems which are utilized by the energy sector entities (e.g., SCADA) ** The use, storage, or handling of various energy materials (oil, coal, gas, etc.) ** The operation and maintenance of such systems
4. Assessment of the likelihood (qualitative probability) of such malevolent acts from adversaries (e.g., terrorists, vandals).	• Determine the possible modes of attack that might result in consequences of significant concern based on critical assets of the Dam Sector entity. The objective of this step of the assessment is to move beyond what is merely possible and determine the likelihood of a particular attack scenario. This is a very difficult task as there is often insufficient information to determine the likelihood of a particular event with any degree of certainty. • The threats (the kind of adversary and the mode of attack) selected for consideration during a vulnerability assessment will dictate, to a great extent, the risk reduction measures that should be designed to counter the threat(s). Some vulnerability assessment methodologies refer to this as a "Design Basis Threat" (DBT) where the threat serves as the basis for the design of countermeasures, as well as the benchmark against which vulnerabilities are assessed. It should be noted that there is no single DBT or threat profile for all hydro or power systems in the United States. Differences in geographic location, size of the utility, previous attacks in the local area and many other factors will influence the threat(s) that the Dam Sector entity should consider in their assessments. Dam Sector entities should consult with the local FBI and/or other law enforcement agencies, public officials, and others to determine the threats upon which their risk reduction measures should be based.
5. Evaluation of existing countermeasures. (Depending on countermeasures already in place, some critical assets may already be sufficiently protect. This step will aid in identification of the areas of greatest concern and help to focus priorities for risk reduction.)	• What capabilities does the system currently employ for detection, delay and response? ** Identify and evaluate current detection capabilities such as intrusion detection systems, energy quality monitoring, operational alarms, guard post orders, and employee security awareness programs. ** Identify current delay mechanisms such as locks and key control, fencing, structure integrity of critical assets and vehicle access checkpoints. ** Identify existing policies and procedures for evaluation and response to intrusion and system malfunction alarms, and system intrusions.

Element	*Points to Consider*
	It is important to determine the performance characteristics. Poorly operated and maintained security technologies provide little or no protection. • What cyber protection system features does the facility have in place? Assess what protective measures are in-place for the SCADA and business-related computer information systems such as: ** Firewalls ** Modem access ** Internet and other external connections, including wireless data and voice communications ** Security polices and protocols **It is important to identify whether vendors have access rights and/or "backdoors" to conduct system diagnostics remotely.** • What security policies and procedures exist, and what is the compliance record for them? Identify existing policies and procedures concerning: ** Personal security ** Physical security ** Key and access badge control ** Control of system configuration and operational data ** Chemical and other vendor deliveries ** Security training and exercise records
6. Analysis of current risk and development of a prioritized plan for risk reduction.	• Information gathered on threat, critical assets, Dam Sector operations, consequences, and and existing countermeasures should be analyzed to determine the current level of risk. The utility should then determine whether current risks are acceptable or risk reduction measures should be pursued. • Recommended actions should measurably reduce risks by reducing vulnerabilities and/or consequences through improved deterrence, delay, detection, and/or response capabilities or by improving operational policies or procedures. Selection of specific risk reduction actions should be completed prior to considering the cost of the recommended action(s). Facilities should carefully consider both short- and long-term solutions. An analysis of the cost of short- and long-term risk reduction actions may impact which actions the utility chooses to achieve its security goals. • Facilities may also want to consider security improvements. Security and general infrastructure may provide significant multiple benefits. For example, improved treatment processes or system redundancies can both reduce vulnerabilities and enhance day-to-day operation.

(*continued*)

Table 5.1. *Continued*

Element	*Points to Consider*
	• Generally, strategies for reducing vulnerabilities fall into three broad categories: ** Sound business practices—affect policies, procedures, and training to improve the overall security-related culture at the facility. For example, it is important to ensure rapid communication capabilities exist between public health authorities and local law enforcement and emergency responders. ** System upgrades—include changes in operations, equipment, processes, or infrastructure itself that make the system fundamentally safer. ** Security upgrades—improve capabilities for detection, delay, or response.

Source: F. R. Spellman (2009). *The Handbook of Safety Engineering: Principles and Applications.* Boca Raton, FL: CRC Press.

- It enables leaders to develop robust methods for managing consequences of asset loss (restoration).
- It provides a means to increase awareness of a broad range of employees to protect truly critical assets and to differentiate in policies and procedures the heightened protection they require.

As previously indicated, identifying the criticality of assets is used primarily to focus the vulnerability analysis efforts. It also assists with the ranking of various recommendations for reducing vulnerabilities. Let's take a look at what electric power infrastructure critical assets might include:

Physical

- Generators
- Substations
- Transformers
- Transmission lines
- Distribution lines
- Control center
- Warehouses
- Office buildings
- Internal and external infrastructure dependencies

Cyber

- SCADA systems
- Networks
- Databases
- Business systems
- Telecommunications

Interdependencies

- Single-point nodes of failures
- Critical infrastructure components of high reliance

VULNERABILITY ASSESSMENT METHODOLOGY (US DOE 2002)

Vulnerability assessment methodology consists of 10 elements. Each element along with a fuller description is listed next.

1. Network architecture
2. Threat environment
3. Penetration testing
4. Physical security
5. Physical asset analysis
6. Operations security
7. Policies and procedures
8. Impact analysis
9. Infrastructure interdependencies
10. Risk characterization

Network Architecture

This element provides an analysis of the information assurance features of the information network(s) associated with the organization's critical information systems. Information examined should include network topology and connectivity (including subnets), principal information assets, interface and communication protocols, function and linage of major software and hardware components (especially those associated with information security such as intrusion detectors), and policies and procedures that govern security features of the network.

Procedures for information assurance in the system, including authentication of access and management of access authorization, should be reviewed. The assessment should identify any obvious concerns related to architectural vulnerabilities, as well as operating procedures. Existing security plans should be evaluated, and the results of any prior testing should be analyzed. Results from the network architecture assessment should include potential recommendations for changes in the information architecture, functional areas and categories where testing is needed, and suggestions regarding system design that would enable more effective information and information system protection.

Three techniques are often used in conducting the network architecture assessment:

1. Analysis of network and system documentation during and after the site visit
2. Interview with facility staff, managers, and chief information officer
3. Tours and physical inspections of key facilities

Threat Environment

Development of a clear understanding of the threat environment is a fundamental element of risk management. When combined with an appreciation of the value of the information assets and systems, and the impact of unauthorized access and subsequent malicious activity, an understanding of threats provides a basis for better defining the level of investment needed to prevent such access.

The threat of a terrorist attack to the Dam Sector infrastructure is real and could come from several areas, including physical, cyber, and interdependency. In addition, threats could come from individuals or organizations motivated by financial gain or persons who derive pleasure from such penetration (e.g., recreational hackers, disgruntled employees). Other possible sources of threats are those who want to accomplish extremist goals (e.g., environmental terrorists, antinuclear advocates) or embarrass one or more organizations.

This element should include a characterization of these and other threats, identification of trends in these threats, and ways in which vulnerabilities are exploited. To the extent possible, characterization of the threat environment should be localized—that is, within the organization's service area.

Penetration Testing

The purpose of network penetration testing is to utilize active scanning and penetration tools to identity vulnerabilities that a determined adversary could easily exploit. Penetration testing can be customized to meet the specific needs and concerns of the Dam Sector unit or utility. In general, penetration testing should include a test plan and details on the rules of engagement (ROD). It should also include a general characterization of the access point sot the critical information systems and include a general characterization of the access points to the critical information systems and communication interface connections, modem network connections, access points to principal network routers, and other external connections. Finally, penetration testing should include identified vulnerabilities and, in particular, whether access could be gained to the control network or specific subsystem or devices that have a critical role in assuring continuity of service.

Penetration testing consists of an overall process of establishing the ground rules or ROE for the test; establishing a white cell for continuous communication; developing a format or methodology for the test; conducting the test; and generating a final report that details methods, findings, and recommendations.

Penetration testing methodology consists of three phases: reconnaissance, scenario development, and exploitation. A one-time penetration test can provide the utility with valuable feedback; however, it is far more effective if performed on a regular basis. Repeated testing is recommended because new threats develop continuously, and the networks, computes, and architecture of the Dam Sector unit or utility are likely to change over time.

Physical Security

The purpose of physical security assessment is to examine and evaluate the systems in place (or being planned) and to identify potential improvements in this area for the sites evaluated. Physical security systems include access controls, barriers locks and keys, badges and passes, intrusion detection devices and associated alarm reporting and display, closed-circuit television (assessment and surveillance), communications equipment (telephone, two-way radio, intercom, cellular), lighting (interior and exterior), power sources (line, battery, generator), inventory control, postings (signs), security system wiring, and protective force (see chapter 9 for greater detail on Dam Sector security devices). Physical security systems are reviewed for design, installation operation, maintenance, and testing.

The physical security assessment should focus on those sites directly related to the critical facilities, including information systems and assets required for operation. Typically included are facilities that house critical equipment or information assets or networks dedicated to the operation of electric, oil, or gas transmission, storage, or delivery systems. Other facilities can be included on the basis of criteria specified by the organization being assessed. Appropriate levels of physical security are contingent upon the value of company assets, the potential threats to these assets, and the cost associated with protecting the assets. Once the cost of implementing/maintaining physical security programs is known, it can be compared to the value of the company assets, thus providing the necessary information for risk management decisions. The focus of the physical security assessment task is determined by prioritizing the company assets; that is, the most critical assets receive the majority of the assessment activity.

At the start of the assessment, survey personnel should develop a prioritized listing of company assets. This list should be discussed with company personnel to identify areas of security strengths and weaknesses. During these initial interviews, assessment area that would provide the most benefit to the company should be identified; once known, they should become the major focus of the assessment activities.

The physical security assessment of each focus area usually consists of the following:

- Physical security program (general)
- Physical security program (planning)
- Barriers
- Access controls/badges
- Locks/keys
- Intrusion detection systems
- Communications equipment
- Protective force/local law enforcement agency

The key to reviewing these topics is not to just identify if they exist but to determine the appropriate level that is necessary and consistent with the value of the asset being protected. The physical security assessment worksheets provide guidance on appropriate levels of protection.

Once the focus and content of the assessment task have been identified, the approach to conduction the assessment can be either at the "implementation level" or at the "organizational level." The approach taken depends on the maturity of the security program.

For example, a company with a solid security infrastructure (staffing plans/procedures, funding) should receive a cursory review of these items; however, facilities where the security programs are being implemented should receive a detailed review. The security staff can act upon deficiencies found at the facilities, once reported.

For facilities with an insufficient security organization, the majority of time spent on the assessment should take place at the organizational level to identify the appropriate staffing/funding necessary to implement security programs to protect assets. Research into specific facility deficiencies should be limited to finding just enough examples to support any staffing/funding recommendations.

Physical Asset Analysis

The purpose of the physical asset analysis is to examine the systems and physical operational assets to ascertain whether vulnerabilities exist. Included in this element is an examination of asset utilization, system redundancies, and emergency operating procedures. Consideration should also be given to the topology and operating practices for electric and gas transmission, processing, storage, and delivery, looking specifically for those elements that either singly or in concert with other factors provide a high potential for disrupting service. This portion of the assessment determines company and industry trends regarding these physical assets. Historic trends, such as asset utilization, maintenance, new infrastructure investments, spare parts, SCADA linkages, and field personnel are part of the scoping element.

The proposed methodology for physical assets is based on a macro-level approach. The analysis can be performed with company data, public data, or both. Some companies might not have readily available data or might be reluctant to share that data.

Key output from analysis should be graphs that show trends. The historic data analysis should be supplemented with on-site interviews and visits. Items to focus on during a site visit include the following:

- Trends in field testing
- Trends in maintenance expenditures
- Trends in infrastructure investments
- Historic infrastructure outages
- Critical system components and potential system bottlenecks
- Overall system operation controls
- Use and dependency of SCADA systems
- Linkages of operation staff with physical and IT security
- Adequate policies and procedures
- Communications with other regional utilities
- Communications with external infrastructure providers
- Adequate organizational structure

Operations Security

Operations security (OPSEC) is the systematic process of denying potential adversaries (including competitors or their agents) information about capabilities and intentions of the host organization. OPSEC involves identifying, controlling, and protecting generally nonsensitive activities concerning planning and execution of sensitive activities. The OPSEC assessment reviews the processes and practices employed for denying adversary access to sensitive and nonsensitive information that might inappropriately aid or abet an individual's or organization's disproportionate influence over system operation (e.g., electric markets, grid operations). This assessment should include a review of security training and awareness programs, discussions with key staff, and tours of appropriate principal facilities. Information that might be available through public access should also be reviewed.

Policies and Procedures

The policies and procedures by which security is administered (1) provide the basis for identifying and resolving issues; (2) establish the standards of reference for policy implementation; and (3) define and communicate roles, responsibilities, authorities, and accountabilities for all individuals and organizations interface with critical systems. They are the backbone for decisions and day-to-day security operations. Security policies and procedures become particularly important at times when multiple parties must interact to effect a desired level of security and when substantial legal ramifications could result from policy violations. Policies and procedures should be reviewed to determine whether they (1) address the key factors affecting security; (2) enable effective compliance, implementation, and enforcement; (3) reference or conform to established standards; (4) provide clear and comprehensive guidance; and (5) effectively address the roles, responsibilities, accountabilities, and authorities.

The objective of the policies and procedures assessment task is to develop a comprehensive understanding of how a facility protects its critical assets through the development and implementation of policies and procedures. Understanding and assessing this area provides a means of identifying strengths and areas for improvements that can be achieved through:

- Modification of current policies and procedures
- Implementation of current policies and procedures
- Development and implementation of new policies and procedures
- Assurance of compliance with policies and procedures
- Cancellation of policies and procedures that are no longer relevant, or are inappropriate, for the facility's current strategy and operations

Impact Analysis

A detailed analysis should be conducted to determine the influence that exploitation of unauthorized access to critical facilities or information systems might have on an organization's operations (e.g., market and/or physical operations). In general, such an analysis would require thorough understanding of (1) the applications and their information processing, (2) decisions influenced by this information, (3) independent checks and balances that might exist regarding information upon which decisions are made, (4) factors that might mitigate the impact of unauthorized access, and (5) secondary impacts of such access (e.g., potential destabilization of organizations serving the grid, particularly those affecting reliability or safety). Similarly, the physical chain of events following disruption, including the primary, secondary, and tertiary impacts of disruption, should be examined.

The purpose of the impact analysis is to help estimate the impact that outages could have on a Dam Sector unit. Outages in electric power, natural gas, and oil can have significant financial and external consequences to a Dam Sector unit. The impact analysis provides an introduction to risk characterization by providing quantitative estimates of these impacts so that the Dam

Sector unit can implement a risk management program and weigh the risks and costs of various mitigation measures.

Infrastructure Interdependencies

The term "infrastructure interdependencies" refers to the physical and electronic (cyber) linkages within and among our nation's critical infrastructures—energy (electric power, oil, natural gas), telecommunications, transportation, water supply systems, banking and finance, emergency services, and government services. This task identifies the direct infrastructure linkages between and among the infrastructures that support critical facilities as recognized by the organization. Performance of this task requires a detailed understanding of an organization's functions, internal infrastructures, and how these link to external infrastructures.

The purpose of the infrastructure interdependencies assessment is to examine and evaluate the infrastructures (internal and external) that support critical facility functions, along with their associated interdependencies and vulnerabilities.

Risk Characterization

Risk characterization provides a framework for prioritizing recommendations across all task areas. The recommendations for each task area are judged against a set of criteria to help prioritize the recommendations and assist the organization in determining the appropriate course of action. It provides a framework for assessing vulnerabilities, threats, and potential impacts (determined in the other tasks). In addition, the existing risk analysis and management process at the organization should be reviewed and, if appropriate, utilized for prioritizing recommendations. The degree to which corporate risk management includes security factors is also evaluated.

VULNERABILITY ASSESSMENT PROCEDURES

Vulnerability assessment procedures can be conducted using various methodologies. For example, the checklist analysis is an effective technology. In addition, Pareto analysis (80/20 principle), relative ranking, pre-removal risk assessment (PRRA), change analysis, failure mode and effects analysis (FMEA), fault tree analysis, event tree analysis, what-if analysis, and Hazard and Operability (HAZOP) can be used in conducting the assessment.

Based on our experience, the what-if analysis and HAZOP seem to be the most user-friendly methodologies to use. A sample what-if analysis procedural outline is presented next, followed by a brief explanation of/and outline for conducting HAZOP.

What-If Analysis Procedure/Sample What-If Questions

The steps in a What-If Checklist analysis are as follows:

1. Select the team (personnel experienced in the process).
2. Assemble information (piping and instrumentation drawings (P&IDs), process flow diagrams (PFDs), operating procedures, equipment drawings, etc.).
3. Develop a list of What-If questions
4. Assemble your team in a room where each team member can view the information
5. Ask each What-If question in turn and determine:
    - What can cause the deviation from design intent that is expressed by the question?
    - What adverse consequences might follow?
    - What are the existing design and procedural safeguards?
    - Are these safeguards adequate?
    - If these safeguards are not adequate, what additional safeguards does the team recommend?
6. As the discussion proceeds, record the answers to these questions in tabular format.
7. Do not restrict yourself to the list of questions that you developed before the project started. The team is free to ask additional questions at any time.
8. When you have finished the What-If questions, proceed to examine the checklist. The purpose of this checklist is to ensure that the team has not forgotten anything. While you are reviewing the checklist, other What-If questions may occur to you.
9. Make sure that you follow up all recommendations and action items that arise from the hazards evaluation.

HAZOP Analysis

The HAZOP analysis technique uses a systematic process to (1) identify possible deviations from normal operations and (2) ensure that safeguards are in place to help prevent accidents. The HAZOP uses special adjectives (such as speed, flow, pressure, etc.) combined process conditions (such as "more," "less," or "no") to systematically consider all credible deviations from normal conditions. The adjectives, called guide words, are a unique feature of HAZOP analysis.

In this approach, each guide word is combined with relevant process parameters and applied at each point (study node, process section, or operating step) in the process that is being examined.

Table 5.2.

Guide Words	*Meaning*
No	Negation of the Design Intent
Less	Quantitative Decrease
More	Quantitative Increase
Part Of	Other Material Present by Intent
As Well As	Other Materials Present Unintentionally
Reverse	Logical Opposite of the Intent
Other Than	Complete Substitution

Table 5.3. Common HAZOP Analysis Process Parameters

Flow	Time	Frequency	Mixing
Pressure	Composition	Viscosity	Addition
Temperature	pH	Voltage	Separation
Level	Speed	Information	Reaction

The following is an example of creating deviations using guide words and process parameters:

Guide Words		*Parameter*		*Deviation*
NO	+	FLOW	=	NO FLOW
MORE	+	PRESSURE	=	HIGH PRESSURE
AS WELL AS	+	ONE PHASE	=	TWO PHASE
OTHER THAN	+	OPERATION	=	MAINTENANCE
MORE	+	LEVEL	=	HIGH LEVEL

Guide words are applied to both the more general parameters (e.g., react, mix) and the more specific parameters (e.g., pressure, temperature). With the general parameters, it is not unusual to have more than one deviation from the application of one guide word. For example, "more reaction" could mean either that a reaction takes place at a faster rate, or that a greater quantity of product results. On the other hand, some combination of guide words and parameters will yield no sensible deviation (e.g., "as well as" with "pressure").

HAZOP Procedure

1. Select the team.
2. Assemble information (P&IDs, PFDs, operating procedures, equipment drawings, etc.).
3. Assemble your team in a room where each team member can view P&IDs.
4. Divide the system you are reviewing into nodes (you can present the nodes, or the team can choose them as you go along).
5. Apply appropriate deviations to each node. For each deviation, address the following questions:
    - What can cause the deviation from design intent?
    - What adverse consequences might follow?
    - What are the existing design and procedural safeguards?
    - Are these safeguards adequate?
    - If these safeguards are not adequate, what does the team recommend?
6. As the discussion proceeds, record the answers to these questions in tabular format.

VULNERABILITY ASSESSMENT: CHECKLIST PROCEDURE

In performing the vulnerability assessment of any Dam Sector unit or facility, one of the simplest methodologies to employ is the checklist. In this section we present a basic outline that is not all inclusive but does contain questions related to each of the infrastructure interdependency survey checklists and their subsections. These sample questions are intended for use by the assessment teams during preparation for interviews with facility representatives to help assure that all relevant aspects of the critical infrastructures are considered in the survey.

Electric Power Supply and Distribution Checklist

Primary Source of Electric Power

- If the primary source of electric power is a commercial source, are there multiple independent feeds? If so, describe the feeds and their locations.
- If the primary source of electric power is a system operated by the facility or asset, what type of system is it?
- If a facility-operated primary electric generation system is used, what are the fuel or fuels used?
- If petroleum fuel is used, what quantity of fuel is stored on site for the primary electric generation system, and how long it would last under different operating conditions?
- If the fuel is stored on site, are arrangements and contracts in place for resupply and management of the fuel?

Electric Distribution System

- Are the components of the electric system that are located outside of buildings (such as generators, fuel storage facilities, transformers, transfer switches) protected from vandalism or accidental damage by fences or barriers? If so, describe the type of protection and level of security it provides.
- Are the various sources of electric power and the components of the internal electric distribution system such that they may be isolated for maintenance or replacement without affecting the critical functions of the asset/facility? If not, describe the limitations.
- Have any single points of failure been identified for the electric power supply and distribution system? If so, list them and describe.

Backup Electric Power Systems

- Are there additional emergency sources of electric supply beyond the primary system (such as multiple independent commercial feeds, backup generators, uninterruptible power supply [UPS])? If there are, describe them.
- If there is a central UPS, does it support all the critical functions of the asset/facility in terms of capacity and connectivity? Specify for how long it can operate on battery power and list any potentially critical functions that are not supported.

- If there is a backup generator system, does it support all the critical functions of the facility in terms of capacity and connectivity? Specify the fuel and list any potentially critical functions that are not supported.
- Is the fuel for the backup generator system a petroleum fuel? If yes, specify the quantity stored on site and how long it would last.
- If the fuel is stored on site, are arrangements and contracts in place for resupply and management of the fuel?

Commercial Electric Power Sources

- How many substations feed the area of the asset facility and the asset/facility itself? That is, is the area supplied by multiple substations? If more than one, which ones have sufficient individual capacities to supply the critical needs of the asset/facility?
- How many distinct independent transmission lines supply the substations?

Commercial Electric Power Pathways

- Are the power lines into the area of the asset/facility and into the asset/facility itself aboveground (on utility poles), buried, or a combination of both? If both, indicate locations of portions aboveground.
- Do the power lines form these substations follow independent pathways to the area of the asset/facility? If not, specify how often and where they intersect or follow the same corridor.
- Are the paths of the power lines co-located with the rights-of-way of other infrastructures? If yes, indicate how often and where they follow the same rights-of-way and the infrastructures that are co-located.
- Are the paths of the power lines located in areas susceptible to natural or accident damage (such as overhead lines near highways; power lines across bridges, dams, or landslide areas)? If yes, indicate the locations and types of potential disruptions.

Commercial Electric Power Contracts

- What type of contract does the asset/facility have with the electric power distribution company or transmission companies? Specify the companies involved and whether there is a direct physical link (distribution or transmission power line) to each company.

- If there is an interruptible contract (even in part), what are the general conditions placed on interruptions such as the minimum quantity that is not interruptible, the maximum number of disruptions per time period, and the maximum duration of disruptions? Has electrical service been interrupted in the past? If yes, describe the circumstances and any effect the outages have had on the critical functions and activities of the asset/facility.

Historical Reliability

- Historically, how reliable has commercial electric power been in the area? Quantify in terms of annual number of disruptions and their durations.
- Typically, when power outages occur, are they of significant duration (as opposed to just a few seconds or minutes)? Quantify the duration of the outages.
- Have there ever been electric power outages of sufficient frequency and duration so as to affect the critical functions and activities of the asset/facility?

Petroleum Fuels Supply and Storage Checklist

Uses of Petroleum Fuels

- Are petroleum fuels used in normal operations at the asset/facility? If yes, specify the steps and uses.
- Are petroleum fuels used during contingency or emergency operations such as for backup equipment or repairs? If yes, specify the types of fuels and their uses.

Reception Facilities

- How are the various petroleum fuels normally delivered to the asset/facility? Indicate the delivery mode and normal frequency of shipments of reach fuel type.
- Under maximum use-rate conditions, are there sufficient reception facilities (truck racks, rail sidings, surge tank capacity, barge moorings) to keep up with maximum contingency or emergency demand)? If no, explain where the expected shortfalls would be and their impacts.
- Are the petroleum fuel delivery pathways co-located with the rights-of-way of other infrastructures or located in areas susceptible to natural or acci-

dental damage (across bridges or dams, in earthquake or landslide areas)? If yes, indicate the locations and types of potential disruptions.
- Are contingency procedures in place to allow for alternative modes or routes of delivery? If yes, describe these alternatives and indicate whether they have sufficient capacity to fully support the critical functions and activities of the asset/facility.

Supply Contracts

- Are contracts in place for the supply of petroleum fuels? Specify the contractors, the types of contracts, the modes of transport (pipeline, rail car, tank truck), and the frequency of normal shipments.
- Are arrangements for emergency deliveries of petroleum fuels in place? Indicate the basic terms of the contracts in terms of the maximum time to delivery and the minimum and maximum quantity per delivery. Also, indicate if these terms as such that there may be effects on the critical functions and activities of the asset/facility.

Natural Gas Supply Checklist

Sources of Natural Gas

- How many city gate stations supply the natural gas distribution system in the area of the asset/facility and the asset facility itself? If more than one, which ones are critical to maintaining the distribution system?
- How many distinct independent transmission pipelines supply the city gate stations? Indicated if an individual gate station is supplied by more than one transmission pipeline and which stations are supplied by independent transmission pipelines.

Pathways of Natural Gas

- Do the distribution pipelines from the individual city gate stations follow independent pathways to the area of the asset/facility? If not, specify how often and where they intersect or follow the same corridor.
- Are the paths of the pipelines co-located with the rights-of-way of other infrastructures? If yes, indicate how often and where they follow the same rights-of-way and the infrastructures that are co-located.

- Are the paths of the pipelines located in areas susceptible to natural or accidental damage (such as across bridges or dams, in earthquake or landslide areas)? If yes, indicate the locations and types of potential disruptions.
- Is the local distribution system will integrated (i.e., can gas readily get from any part of the system to any other part of the system)?

Natural Gas Contracts

- Does the asset/facility have a firm delivery contract, an interruptible contract, or a mixed contract with the natural gas distribution company or the transmission companies? Specify the companies involved and whether there is a direct physical ling (pipeline) to each company.
- If there is an uninterruptible contract (even in part), what are the general conditions placed on interruptions such as the minimum quantity that is into interruptible, the maximum number of disruptions per time period, and the maximum duration of disruptions? Has natural gas service been interrupted in the past? If yes, describe the circumstances and any effect the outages have had on the critical functions and activities of the asset/facility.
- Does the asset/facility have storage or some other sort of special contracts with natural gas transmission or storage companies? If yes, briefly describe the effect on sustaining a continuous supply of natural gas to the asset/facility.
- In case of a prolonged disruption of natural gas supply, are contingency procedures in place to allow for the use of alternative fuels (such as on-site propane-air, liquefied petroleum gas, or petroleum fuels)? If yes, describe these alternatives and indicate whether they have sufficient capacity to fully support the critical functions and activities of the asset/facility.

Historical Reliability

- Historically, how reliable has the natural gas supply been in the area? Quantify by describing any unscheduled or unexpected disruptions. Were there any effects on the critical functions and activities of the asset/facility?
- If operating under an uninterruptible service agreement, has natural gas service ever been curtailed? If yes, how often, for how long, and were there any effects on the critical functions and activities of the asset/facility?

REFERENCES AND RECOMMENDED READING

Belke, J. C. 2001. "Chemical Accident Risks in U.S. Industry—A Preliminary Analysis of Accident Risk Data from U.S. Hazardous Chemical Facilities." *Proceedings of the 10th International Symposium on Loss Prevention and Safety Promotion in the Process Industries*, Stockholm, Sweden. Amsterdam: Elsevier Science.

CBO. 2004. *Homeland Security and the Private Sector*. www.cbo.gov/ft. docs (accessed May 2, 2008).

Clark, R. M., and R. A. Deininger. 2000. "Protecting the Nation's Critical Infrastructure: The Vulnerability of U.S. Water Supply Systems." *Journal of Contingencies and Crisis Management* 8 (2): 76–80.

CSI. 2011. *2010/2011 Computer Crime and Security Survey*. Orlando, FL: Computer Security Institute.

DHS. 2009. "National Infrastructure Protection Plan." http://www.dhs.gov/xlibrary/assets/NIPP.Plan.pdf (accessed May 1, 2009).

FBI. 2009. "Terrorism Threats to Energy Sector; Congressional Testimony of FBI Director." http://www.fib.gov/congress/congress03/muller021103.htm (accessed May 1, 2009).

FBI. 2012. "Insider Espionage. Report before the House Committee on Homeland Security Subcommittee on Counter Terrorism and Intelligence." Washington, DC: Federal Bureau of Investigation.

FEMA. 2008. *FEMA452: Risk Assessment A How to Guide*. fema.gov/library/file?type=published/filetofile (accessed May 1, 2008).

FEMA. 2015. *Protecting Critical Infrastructure Against Insider Threats*. http://emilms.fema.gov/IS0915/IABsummary.htm (accessed April 17, 2015).

National Infrastructure Advisory Council. 2008. *First Report and Recommendations on the Insider Threat to Critical Infrastructure*. Washington, DC.

Spellman, F. R. 1997. *A Guide to Compliance for PSM/RMP*. Lancaster, PA: Technomic Publishing Company.

US DOE. 2010. *Energy Sector-Specific Plan: An Annex to the National Infrastructure Protection Plan*. Washington, DC.

US DOE. 2002. *Vulnerability Assessment Methodology: Electric Power Infrastructure*. Washington, DC.

US EPA. 2002. *Vulnerability Assessment Fact Sheet*. EPA 816-F-02-025. www.epa.gov/ogwdw/security/index.html (accessed May 2006).

6

Preparation

When Is Enough, Enough?

Fighting terrorism is like being a goalkeeper. You can make a hundred brilliant saves but the only shot that people remember is the one that gets past you.

—Paul Wilkinson (1992)

We must take the battle to the enemy, disrupt his plans, and confront the worst threats before they emerge.

—George W. Bush

Americans should find comfort in knowing that millions of their fellow citizens are working every day to ensure our security at every level—federal, state, county, municipal. These are dedicated professionals who are good at what they do. I've seen it up close, as Governor of Pennsylvania. . . . But there may be gaps in the system. The job of the Office of Homeland Security will be to identify those gaps and work to close them.

Now, obviously, the further removed we get from September 11, I think the natural tendency is to let down our guard. Unfortunately, we cannot do that. The government will continue to do everything we can to find and stop those who seek to harm us. And I believe we owe it to the American people to remind them that they must be vigilant, as well.

—Tom Ridge (quoted in Henry, 2002)

The possibility of dam infrastructure terrorism—attacks on the U.S. dam infrastructure—doesn't generate the same attention as potential nuclear, biological, or chemical terrorism. However, because of the seriousness of the threat of terrorism to the our hydropower, navigation, and levee sector and the enormous economic implications of such attacks, USDOE, USEPA, and other agencies have worked nonstop since 9/11 in gathering and providing as much advice and guidance as possible to aid dam sector personnel in protecting dam sector production and protection facilities and associated critical support infrastructure. This chapter provides an overview of important tools that can be used in protecting dam sector facilities; that is, conventional dam or retaining structures, reservoirs, spillways, outlet works, powerhouses, and canals/aqueducts.

DAM SECTOR THREATS AND INCIDENTS

Based on evidence of potential losses from past natural and human-made accidents, indication of the potential human and environmental losses and economic costs from an attack on a large dam facility comes from major accidents that have occurred both abroad and in the United States. Those events indicate that the human and environmental losses could be significant (CBO 2004).

Dam Sector threats and incidents may be of particular concern due to the range of potential consequences:

- Disrupting system operations and thus interrupting the supply of electrical energy and ease of navigation.
- Causing physical damage to system infrastructure.
- Reducing public confidence in the hydropower energy supply system.
- Long-term denial of electrical energy and the cost of replacement.
- Creating an adverse impact on public health within a population.

Keep in mind that some of these consequences would only be realized in the event of a successful terrorist incident; however, the mere *threat* of terrorism can also have an adverse impact on industries that depend on a safe, steady supply of energy. In addition, the economic implications of such attacks are potentially enormous.

While it is important to consider the range of possibilities associated with a particular threat, assessments are typically based on the probability of a particular occurrence. Determining probability is somewhat subjective, and is often based on intelligence and previous incidents. As mentioned, there are historical accounts of accidental incidents that have caused tremendous death and destruction.

THREAT WARNING SIGNS[1]

A threat warning is an occurrence or discovery that indicates a potential threat that triggers an evaluation of the threat. It is important to note that these warnings must be evaluated in the context of typical industry activity and previous experience in order to avoid false alarms. Following is a brief description of potential warnings.

- **Security Breach**. Physical security breaches, such as unsecured doors, open hatches, and unlocked/forced gates, are probably the most common threat warnings. In most cases, the security breach is likely related to lax operations or typical criminal activity such as trespassing, vandalism, and theft. However, it is wise to assess *any* security breach with respect to the possibility of attack.
- **Witness Account**. Awareness of an incident may be triggered by a witness account of tampering. Dam Sector sites/facilities should be aware that individuals observing suspicious behavior near Dam Sector facilities/plants will likely call 911 and not the plant. In this case, the incident warning technically might come from law enforcement. Note: the witness may be a plant employee engaged in their normal duties.
- **Direct Notification by Perpetrator**. A threat may be made directly to the dam sector site, plant or facility, either verbally or in writing, including via Facebook and by Twitter. Historical incidents would indicate that verbal threats made over the phone are more likely than written threats. While the notification may be a hoax, threatening a Dam Sector unit is a crime and should be taken seriously.
- **Notification by Law Enforcement**. A Dam Sector site/facility may receive notification about a threat directly from law enforcement, including local, country, state, or federal agencies. As discussed previously, such a threat could be a result of suspicious activity reported to law enforcement, either by a perpetrator, a witness, or the news media. Other information, gathered through intelligence or informants, could also lead law enforce-

1. The following section is from USEPA (2003), *Response Protocol Toolbox: Planning for and Responding to Drinking Water Contamination Threats and Incidents* (Washington, DC: United States Environmental Protection Agency).

ment to conclude that there may be a threat to the Dam Sector site/facility. While law enforcement will have to be lead in the criminal investigation, the dam site/facility has primary responsibility for the safety of it equipment, electricity-producing processes, and public health. Thus, the plant's role will likely be to help law enforcement to appreciate the public health implications of a particular threat as well as the technical feasibility of carrying out a particular threat.

- **Notification by News Media**. A threat to destroy a Dam Sector site/facility might be delivered to the news media or the media may discover a threat. A conscientious journalist would immediately report such a threat to the police, and either the journalist or the police would immediately contact the dam site/facility. This level of professionalism would provide an opportunity for the plant to work with the media and law enforcement to assess the credibility of the threat before any broader notification is made.
- **Public Health Notification**. In this case, the first indication that electrical blackouts or dam sector site/facility emergencies (e.g., sabotage) has occurred is the appearance of victims in local emergency rooms and health clinics. Dam sites/facilities may therefore be notified, particularly if the cause is unknown or linked to electricity production, distribution, or transmission. An incident triggered by a public health notification is unique in that at least a segment of the population has been exposed to a harmful substance. If this agent is a hazardous petrochemical triggered by electrical power failure or interruption, then the time between exposure and onset of symptoms may be on the order of hours, and thus there is the potential that the contaminant is still present.

RESPONSE TO THREATS: WHEN IS ENOUGH, ENOUGH?

This section is *not* designed to discuss what specific steps to take in responding to a terrorist threat. Rather, the questions addressed in this section are: "Why is it necessary to plan to respond to Dam Sector threats at all?" and "When have I done enough?"

Federal, state, and local programs already exist that—with varying degrees of effectiveness—encourage or require the operators of Dam Sector sites/facilities to boost their efforts to promote safety and security and to share information that can help local governments plan for emergencies (Schierow 2004).

Proper planning is a delicate process because public health measures are rarely noticed or appreciated (like buried utility pipes, they are often hidden functions) except when they fail—then they are very visible. Consumers are particularly upset by unreliable dam-produced electrical supply systems and energy products or practices that produce unsafe (contaminated) environmental media—water, air, and/or soil—because they are often viewed as entitlements, and, indeed, it is reasonable for consumers to expect a high-quality, safe environment. Public health failures during a response to contamination threats often take the form of too much or too little action. The results of too little action, including no response at all, can have disastrous consequences potentially resulting in public injuries or fatalities. On the other hand, a disproportionate response to hazardous materials contamination threats that have not been corroborated (i.e., determined to be "credible") can also have serious repercussions: over-response to threats carries its own adverse impacts. Consider, for example, if otherwise safe drinking water is unavailable because of an uncorroborated threat that it has been contaminated with hazardous chemicals or materials resulting from electrical failure or interruption. Not only would the water be unavailable for human consumption, but it would also be unavailable for sanitation, firefighting, industry, and the many other typical uses of public water supply.

Although precise estimates are not available, information on accidents in the Dam Sector, company assessment of what could happen in a severe release or explosion, and actual terrorist incidents involving destruction of or interruption of electricity supplies suggest that the risk of attack is real and the losses could be significant. These adverse impacts must be considered when evaluating response options to a Dam Sector threat.

Considering the potential risks of an inappropriate response to a severe water release or explosion threat, it is clear that a systematic approach is needed to evaluate Dam Sector threats. One overriding question is "When has a Dam Sector entity done enough?" This question may be particularly difficult to address when considering the wide range of agencies that may be involved in a threat situation. Other organizations, such as USEPA, USDOE, CDC, USDOT, USGS, law enforcement agencies, public health departments, and so on, will each have unique obligations or interests in responding to a severe water release or explosion threat.

The guiding principle for responding to severe water release and/or destruction of navigation asset threats is one of "due diligence" or "what is a suitable and sensible response to a threat?" As discussed previously, some response to Dam Sector threats is warranted due to the public health implications of an actual sector destruction incident. However, a Dam Sector unit could spend a lot of time and money over-responding to every threat, which would be an ineffective use of resources.

Ultimately, the answer to the question of "due diligence" must be decided at the local level and will depend on a number of considerations. Among other factors, local authorities must decide what level of risk is reasonable in the context of a perceived threat. Careful planning is essential to developing an appropriate response to terrorist threats, and in fact, one primary objective of USEPA's *Response Protocol Tool Box* (RPTB) is to aid users in the development of their own site-specific plans that are consistent with the needs and responsibilities of the user. Beyond planning, the RPTB considers a careful evaluation of any terrorist threat, and an appropriate response based on the evaluation, to be the most important element of due diligence.

In the RPTB, the threat management process is considered in three successive stages: **possible**, **credible**, and **confirmed**. Thus, as the threat escalates through these three states, the actions that might be considered due diligence expand accordingly. The following paragraphs describe, in general terms, actions that might be considered as due diligence at these various stages.

- **Stage 1: "Is the threat possible?"** If a Dam Sector facility is faced with a terrorism threat, they should evaluate the available information to determine whether or not the threat is "possible" (i.e., could something have actually happened). If the threat is possible, immediate operational response actions might be implemented, and activities such as site characterization would be initiated to collect additional information to support the next stage of the threat evaluation.
- **Stage 2: "Is the threat credible?"** Once a threat is considered "possible," additional information will be necessary to determine if the threat is "credible." The threshold at the credible stage is higher than that at the possible stage, and in general there must be information to corroborate the threat in order for it to be considered credible.

- **Stage 3: "Has the incident been confirmed?"** Confirmation implies that definitive evidence and information have been collected to establish the presence of a threat to the Dam Sector. Obviously, at this stage the concept of due diligence takes on a whole new meaning since authorities are now faced with death and destruction and a potential public health crisis. Response actions at this point include all steps necessary to protect public health, property and the environment.

PROTECTIVE MEASURES AND RESILIENCE STRATEGIES

Later in this book we provide a detailed discussion of security techniques and hardware. However, at this point it is important to outline basic protective measures and resilience strategies—because in the Dam Sector these measures and strategies will vary depending on asset-specific characteristics and will consist of physical items and systems as well as plans to address contingencies and respond to incidents. *The Dams Sector Protective Measures Handbook* provides a comprehensive list of risk reduction strategies and protective measures that could be implemented at sector assets. The following are examples of general protective measures.

Access Control

- Various levels of access control can be used depending on the critical nature of the area being accessed. Electronic entry control systems supplemented by CCTV and adequate lighting might be used for personnel entering areas of critical importance.
- Intrusion detection systems may also be activated at portals.
- Effective locking devices may be installed at access points to gate/intake structures, galleries, vaults, and other critical locations.

Barriers

- Perimeter barriers (e.g., fences, berms, vehicle-resistant barriers, and boat barriers) could be installed to maintain a clear area at the perimeter, enable monitoring, and incorporate adequate standoff distances.
- Intrusion barriers might be used at outlet conduits.

Monitoring and Surveillance

- CCTV cameras facilitate passive monitoring (e.g., review of taped incident after it occurs) and active monitoring (e.g., viewing of real-time activities).

- Detector and alarm systems (e.g., intrusion detectors, fire and smoke alarms, and motion detectors) might be integrated with surveillance systems to promote immediate assessment of any detected intrusion.

Planning and Preparation

- Emergency action plans
- Site security plans and threat response procedures
- Recovery plans and continuity-of-operation plans
- Exercises to test the effectiveness of these plans and procedures

Personnel and Site Operations

- Pre-employment criminal background checks may be conducted for new hires, seasonal employees, and temporary agency personnel.
- Vendors, contractors, and service providers with access to critical areas and information could be required to verify that their employees have been subjected to a pre-employment criminal background check.
- All Dam Sector managers and process equipment operating personnel must know their facilities. For these persons, there is no excuse for not knowing every square inch of the site. In particular, site workers should know about any and all construction activities underway on the site; the actual construction parameters of the site; and especially operation of all plant unit processes. In addition, site management must not only know their operating staff but also their major customers.

Communication

- Reliable communication systems (primary and alternate) may be established to provide effect channels to reach local law enforcement and emergency responders.
- Notification protocols may be developed that outline who to contact in the event of an incident, providing staff with training and checklists.

Cybersecurity

- Security plans may be developed and implemented for computer and information systems hardware and software to provide for more secure computer network architectures.

- Computer systems recovery and restoration plans may be prepared to return computer systems to full functionality after an incident.
- Up-to-date cybersecurity techniques (e.g., firewalls, virus protection, anti-spyware protection, encryption, and user authentication) and software patches may be used on commercial facility computer systems. Control systems can be isolated from the Internet where it is practical to do so.
- Redundancies in computer hardware and software can be used to permit the continued operation of information systems should the primary systems become disabled.
- Redundancies for control systems could also include the use of operators and manual controls for the system in the event of disruption. Off-site backup facilities can ensure the operation of control systems in the event of total failure at the primary sites.
- Procedural and managerial requirements may be developed to include changing control and firewall monitoring, firewall assurance, standards, and training.

Redundancy

- Redundancy and emergency backup capabilities can be implemented to provide critical utility services (e.g., backup electric power generators and multiple utility feeder lines). When feasible, redundant and backup equipment may be stored in a different part of the facility than the primary supply equipment.

Structural Retrofit

- Harden, fortify, or modify the asset to reduce its vulnerability to attack or the likelihood of failure from a natural event.

REFERENCES AND RECOMMENDED READING

AWWARF. 2002. "Online Monitoring for Drinking Water Utilities." Erika Hargesheimer (ed.), AWWA Research Foundation, and CRS PORAQUA. Denver, CO: American Water Works Association.

CBO. 2004. *Homeland Security and the Private Sector*. Washington, DC.

Henry, K. 2002. "New Face of Security." *Government Security*, 30–37.

Schierow, L. 2004. "CRS Report for Congress: Chemical Plant Security." Congressional Research Center, U.S. Library of Congress.

USDHS. 2007. *Dams Protective Measures and Resilience Handbook*. Washington, DC: U.S. Department of Homeland Security.

USEPA. 2001. *Protecting the Nation's Water Supplies from Terrorist Attack: Frequently Asked Questions*. Washington, DC: United States Environmental Protection Agency.

USEPA. 2003. *Response Protocol Toolbox: Planning for and Responding to Drinking Water Contamination Threats and Incidents*. Washington, DC: United States Environmental Protection Agency.

WHO. 2003. *Public Health Response to Biological and Chemical Weapons: WHO Guidance, 2nd ed*. Geneva, Switzerland.

7

Cybersecurity and Dam Sector Power Control Systems

Terrorism is the price of empire. If you do not wish to pay the price, you must give up the empire.

—Pat Buchanan

Unless people are injured, there is also less drama and emotional appeal.

—Dorothy Denning, Fellow of the Association for Computing Machinery, "Is Cyber Terror Next?"

A "war against terrorism" is an impracticable conception if it means fighting terrorism with terrorism.

—John Mortimer, *Thoughts of the Good Life*

In Queensland, Australia, on April 23, 2000, police stopped a car on the road to Deception Bay and found a stolen computer and radio transmitter inside. Using commercially available technology, Vitek Boden, 48, had turned his vehicle into a pirate command center for sewage treatment along Australia's Sunshine Coast.

Boden's arrest solved a mystery that had troubled the Maroochy Shire wastewater system for two months. Somehow the system was leaking hundreds of thousands of gallons of putrid sludge into parks, rivers and the manicured grounds of a Hyatt Regency hotel. . . . Until Boden's capture—during his 46th successful intrusion—the utility's managers did not know why.

—Barton Gellman, "Cyber-Attacks by Al Qaeda Feared," *Washington Post*

THE LANGUAGE OF CYBERSECURITY

In 2000, the FBI identified and listed threats to critical infrastructure. These threats are listed and described in table 7.1. What is of course striking is the prevalence of cybersecurity threats—something new then. Today, these kinds of threats and attacks are sadly no longer unusual or unexpected.

Today's fast-evolving technology has intensified the importance of critical infrastructure protection, in which cybersecurity has become as critical as physical security to protecting Dam Sector critical infrastructure.

In the past decade, it has been somewhat routine for us to see a television news program or read an online media exposé where the focus is on cyberse-

Table 7.1. Threats to Critical Infrastructure Observed by the FBI

Threat	*Description*
Criminal groups	There is an increased use of cyber intrusions by criminal groups who attack systems for purposes of monetary gain.
Foreign intelligence services	Foreign intelligence services use cyber tools as part of their information gathering and espionage activities.
Hackers	Hackers sometimes crack into networks for the thrill of the challenge or for bragging rights in the hacker community. While remote cracking once required a fair amount of skill or computer knowledge, hackers can now download attack scripts and protocols from the Internet and launch them against victim sites. Thus, while attack tools have become more sophisticated, they have also become easier to use.
Hacktivists	Hacktivism refers to politically motivated attacks on publicly accessible Web pages or e-mail servers. These groups and individuals overload e-mail servers and hack into websites to send a political message.
Information warfare	Several nations are aggressively working to develop information warfare doctrine, programs, and capabilities. Such capabilities enable a single entity to have a significant and serious impact by disrupting the supply, communications, and economic infrastructures that support military power—impacts that, according to the Director of Central Intelligence, can affect the daily lives of Americans across the country.
Inside threat	The disgruntled organization insider is a principal source of computer crimes. Insiders may not need a great deal of knowledge about computer intrusions because their knowledge of a victim system often allows them to gain unrestricted access to cause damage to the system or to steal system data. The insider threat also includes outsourcing vendors.
Virus writers	Virus writers are posing an increasingly serious threat. Several destructive computer viruses and "worms" have harmed files and hard drives, including the Melissa Macro Virus, the Explore. Zip worm, the CIH (Chernobyl) Virus, Nimda, and Code Red.

Source: FBI, 2000.

curity—or, more often, the *lack* of it. When we read or hear about cyber intrusion incidents, the reports add new terms or new uses for old terms to our vocabulary. For example, old terms such as botnets (short for robot networks, aka balled bots, zombies, botnet fleets, etc.)—groups of computers that have been compromised with malware such as Trojan Horses, worms, backdoors, remote control software, and viruses—have taken on new connotations with regard to cybersecurity. Relatively new terms such as scanners, Windows NT hacking tools, ICQ hacking tools, mail bombs, sniffer, logic bomb, nukers, dots, backdoor Trojan, key loggers, hackers' Swiss knife, password crackers, blended threats, Warhol Worms, Flash Threats, Targeted Attacks, and BIOS crackers are now common. And new terms have evolved along with various control mechanisms. For example, because many control systems are vulnerable to attacks of varying degrees, these attack attempts range from telephone line sweeps (wardialing), to wireless network sniffing (wardriving), to physical network port scanning, and to physical monitoring and intrusion. When wireless network sniffing is performed at (or near) the target point by a pedestrian, that is called warwalking, and it means that instead of a person being in a vehicle, the potential intruder may be sniffing the network for weaknesses or vulnerabilities on foot, posing as a pedestrian, using a handheld PDA device or laptop computer (Warwalking 2003).

MASTER SUN TZU, FROM *THE ART OF WAR*

Those who are first on the battlefield, and await the opponents are at ease; those who are last, and head into battle are worn out.

Not all relatively new and universally recognizable cyber terms have sinister connotations or meanings, of course. Consider, for example, the following digital terms: backup, binary, bit byte, CPU, database, e-mail, HTML, icon, memory, cyberspace, modem, monitor, network, RAM, Wi-Fi (wireless fidelity), record, software, World Wide Web. None of these terms normally conjure up images of terrorism.

There is, however, one digital term, SCADA, that most people have not heard of. This is not the case, however, with those who work with the nation's critical infrastructure, including Dam Sector infrastructure. SCADA, or **S**upervisory **C**ontrol **a**nd **D**ata **A**cquisition System (also sometimes referred to as Digital Control Systems or Process Control Systems), plays an important role in computer-based control systems. From coordinating music and lights in the proper sequence with spray from water fountains to controlling systems used in the drilling and refining of oil and natural gas, to delivering extreme availability, fault tolerance, performance, security, and wide-area communications; indeed, power control systems perform many functions. Many Dam Sector distribution networks use computer-based systems to remotely control sensitive feeds, processes and system equipment previously controlled manually. These systems allow a Dam Sector entity (or operation) to monitor power availability levels, to ensure that electrical power is produced at optimum levels, and, as mentioned, collect data from sensors and control equipment located at remote sites. Common power control systems for many critical infrastructures measure elements such as fluid level, temperature, pressure, chemical purity or slurry composition, chemical purity, and energy and electrical distribution flow rates. Common Dam Sector power control system equipment includes valves, pumps, and switching devices for distribution of electricity.

The critical infrastructure of many countries is increasingly dependent on SCADA systems. There is a newer power control system now installed at Grand Coulee Dam (see figures 7.1 and 7.2) known as RTI (Real Time Innovations) Connext (TM) DDS. Future plans call for additional installations at other Dam Sector facilities.

The installation of the RTI power control system at Grand Coulee Dam was a daunting challenge. This was the case, as Carrieann Stocks (2014) points out, because the power control system at Grand Coulee Dam is the most challenging and mission-control system in the power industry; however, the RTI middleware successfully implements the facility's communication, module failover, wide-area routing, control room integration, and operator interface. RTI has proven itself capable of handling all of Grand Coulee Dam's performance, scale, reliability, and scalability requirements.

FIGURE 7.1
Grand Coulee Dam. Photo by F. R. Spellman.

FIGURE 7.2
Grand Coulee Dam. Photo by F. R. Spellman.

RTI and systems like it are undoubtedly the power control systems of the future. At the present time, however, Grand Coulee Dam is the only system fully employing the use of RTI and other Dam Sectors facilities are to be retrofitted with the system on an ongoing basis in the future. Presently, SCADA is the system that is most used in the other Dam Sector facilities. Therefore, in this text the major focus in on SCADA.

WHAT IS SCADA?

If we were to ask the specialist to define SCADA, the technical response could be outlined as follows:

SCADA

- is a multitier system or interfaces with multitier systems
- is used for physical measurement and control endpoints via an RTU and PLC to measure voltage, adjust a value or flip a switch
- is an intermediate processor normally based on commercial third-party OSes—VMS, Unix, Windows, Linux
- is human interfaces, for example, with graphical user interface (Windows GUIs)
- has a communication infrastructure that consists of a variety of transport mediums such as analog, serial, Internet, radio, Wi-Fi

But how about the nonspecialist response?—for the rest of us who are nonspecialists? Well, for the nonspecialist and the rest of us we could simply say that SCADA is a computer-based control system that remotely controls processes previously controlled manually. The philosophy behind SCADA control systems can be summed up by the phrase "If you can measure it, you can control it." SCADA allows an operator using a central computer to supervise (control and monitor) multiple networked computers at remote locations. Each remote computer can control mechanical processes (mixers, pumps, valves, etc.) and collect data from sensors at its remote location. Thus the phrase: **S**upervisory **C**ontrol **a**nd **D**ata **A**cquisition, or **SCADA**.

The central computer is called the Master Terminal Unit, or MTU. The MTU has two main functions: periodically obtain data from RTUs/PLCs and control remote devices through the operator station. The operator interfaces with the MTU using software called Human Machine Interface, or HMI. The remote computer is called Program Logic Controller (PLC) or Remote Terminal Unit (RTU). The RTU activates a relay (or switch) that turns mechanical equipment "on" and "off." The RTU also collects data from sensors. Sensors perform measurement, and actuators perform control.

In the initial stages utilities ran wires, also known as hardwire or land lines, from the central computer (MTU) to the remote computers (RTUs). Because remote sites can be located hundreds of miles from the central location utilities begun to use public phone lines and modems, leased telephone company lines, and radio and microwave communication. More recently, they have also begun to use satellite links, Internet, and newly developed wireless technologies.

DID YOU KNOW?

Modern RTUs typically use a ladder-logic approach to programming due to its similarity to standard electrical circuits. A RTU that employs this ladder logic programming is called a Programmable Logic Controller (PLC).

Because SCADA systems' sensors provided valuable information, many utilities and other industries established "connections" between their SCADA systems and their business system. This allowed Utility/Industrial management and other staff access to valuable statistics, such as chemical usage. When utilities/industries later connected their systems to the Internet, they were able to provide stakeholders/stockholders with usage statistics on the Utility/Industrial web pages. Figure 7.3 provides a basic illustration of a representative SCADA network. Note that firewall protection (see chapter 9) would normally be placed between Internet and business systems and between business systems and the MTU.

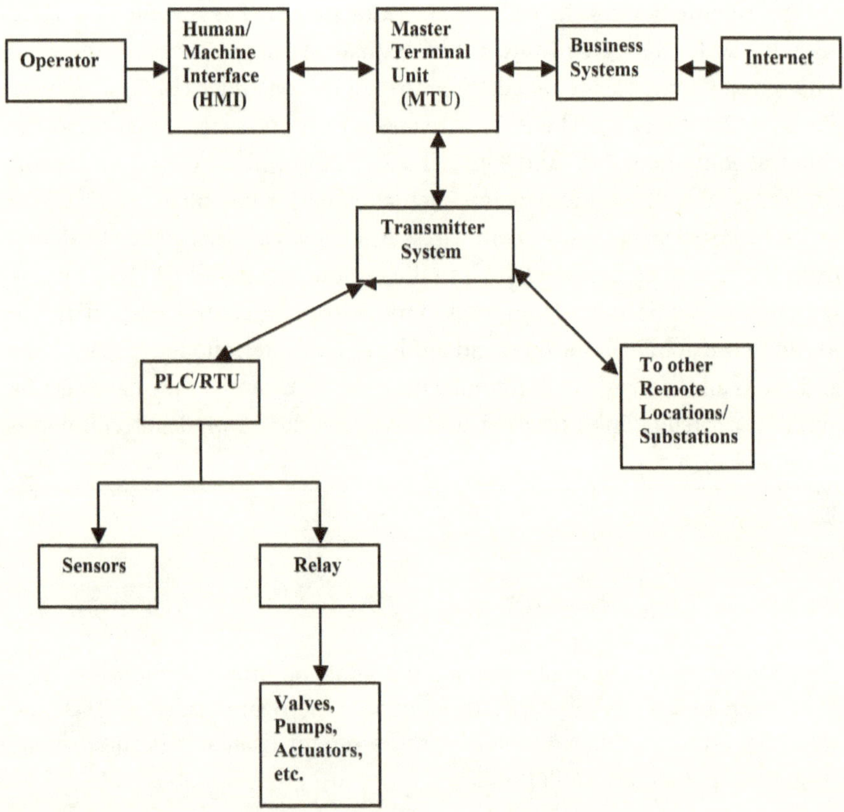

FIGURE 7.3
Representative SCADA network.

SCADA APPLICATIONS

As stated previously, SCADA systems can be designed to measure a variety of equipment-operating conditions; to measure parameters or volumes and flow rates of electricity, natural gas, and oil or oil and petrochemical mixture quality parameters; and to respond to changes in those parameters either by alerting operators or by modifying system operation through a feedback loop system—without having personnel physically visit each process or piece of equipment on a daily basis to check it and/or ensure that it is functioning properly. Automation and integration of large scale diverse assets require SCADA systems to provide the utmost in flexibility, scalability, openness, and reliability. SCADA systems are used to automate certain energy production functions; these can be performed without initiation by an operator. In addition to process equipment, SCADA systems can also integrate specific security alarms and equipment, such as cameras, motion sensors, lights, data from card reading systems, etc., thereby providing a clear picture of what is happening at areas throughout a facility. Finally, SCADA systems also provide constant, real-time data on processes, equipment, location access, and so on, the necessary response to be made quickly.

To move from the general to the specific, common SCADA systems applications for the Dam Sector include, but are not limited to, these shown here.

- Bearing temperature monitor (electric generators and motors)
- Gas processing
- Plant energy management
- Power distribution monitoring
- Electric power monitoring
- Fuel oil handling system
- Hydroelectric load management
- Plant monitoring
- Process controls
- Process stimulators
- Tank controls
- Utility monitoring
- Safety parameter display systems
- Turbine controls
- Turbine monitoring
- Virtual annunciator panels

Because these systems can monitor multiple processes, equipment, and infrastructure and then provide quick notification of, or response to, problems or upsets, SCADA systems typically provide the first line of detection for atypical or abnormal conditions. For example, a SCADA system connected to sensors that measure specific Dam Sector quality parameters are measured outside of a specific range. A real-time customized operator interface screen could display and control critical systems monitoring parameters.

The system could transmit warning signals back to the operators, such as by initiating a call to a personal pager. This might allow the operators to initiate actions to prevent power outages or contamination and disruption of the electrical energy supply. Further automation of the system could ensure that the system initiated measures to rectify the problem. Preprogrammed control functions (e.g., shutting a valve, controlling flow, throwing a switch, or adding chemicals) can be triggered and operated based on SCADA utility.

SCADA VULNERABILITIES

U.S. ELECTRIC GRID IS HACKED

The Associated Press (AP) reported on April 9, 2009, that spies hacked into the U.S. energy grid and left behind computer programs (Trojan Horses) that would enable them to disrupt service, exposing potentially catastrophic vulnerabilities in key pieces of national infrastructure.

Even though terrorists, domestic and/or foreign, tend to aim their main focus around the critical devices that control actual Dam Sector energy production and delivery, and/or operation according to USEPA (2005), SCADA networks were originally developed with little attention given to security, making the security of these systems often weak. Studies have found that, while technological advancements introduced vulnerabilities, many Dam Sector plants/sites and utilities have spent little time securing their SCADA networks. As a result, many SCADA networks may be susceptible to attacks and misuse. SCADA systems languished in obscurity and this was the essence of their security; that is, until technological developments transformed SCADA from a backroom operation to a front-and-center visible control system.

Remote monitoring and supervisory control of processes begun to develop in the early 1960s, and adopted many technological advancements. The advent of minicomputers made it possible to automate a vast number of once manually operated switches. Advancements in radio technology reduced the communication costs associated with installing and maintaining buried cable in remote areas. SCADA systems continued to adopt new communication methods including satellite and cellular. As the price of computers and communications dropped, it became economically feasible to distribute operations and to expand SCADA networks to include even smaller facilities.

Advances in information technology and the necessity of improved efficiency have resulted in increasingly automated and interlinked infrastructures and have created new vulnerabilities due to equipment failure, human error, weather and other natural causes, and physical and cyberattacks. Some areas and examples of possible SCADA vulnerabilities include:

- **Human**—People can be tricked or corrupted, and may commit errors.
- **Communications**—Messages can be fabricated, intercepted, changed, deleted, or blocked.
- **Hardware**—Security features are not easily adapted to small self-contained units with limited power supplies.
- **Physical**—Intruders can break into a facility to steal or damage SCADA equipment.
- **Natural**—Tornadoes, floods, earthquakes, and other natural disasters can damage equipment and connections.
- **Software**—Programs can be poorly written.

A 1998 study by B. C. Ezell included a survey that found that many water utilities were doing little to secure their SCADA network vulnerabilities (Ezell 1998). For example, many respondents reported that they had remote access, which could allow an unauthorized person to access the system without being physically present. More than 60 percent of the respondents believed that their systems were not safe from unauthorized access and use. Twenty percent of the respondents even reported known attempts, successful unauthorized access, or use of their system. Yet 22 of 43 respondents reported that they do not spend any time ensuring their network is safe and 18 of 43 respondents reported that they spend less than 10 percent ensuring network safety.

SCADA system computers and their connections are susceptible to different types of information system attacks and misuse such as system penetration and unauthorized access to information. The Computer Security Institute and Federal Bureau of Investigation conduct an annual Computer Crime and Security Survey (FBI 2004). The survey reported on 10 types of attacks or misuse and reported that virus and denial of service had the greatest negative economic impact. The same study also found that fifteen percent of the respondents reported abuse of wireless networks, which can be a SCADA component. On average, respondents from all sectors did not believe that their organization invested enough in security awareness. Utilities as a group reported a lower average computer security expenditure/investment per employee than many other sectors such as transportation telecommunications, and financial.

Sandia National Laboratories' *Common Vulnerabilities in Critical Infrastructure Control Systems* described some of the common problems it has identified in the following five categories (Stamp et al. 2003):

1. **System Data**—Important data attributes for security include availability, authenticity, integrity, and confidentiality. Data should be categorized according to its sensitivity, and ownership and responsibility must be assigned. However, SCADA data is often not classified at all, making it difficult to identify where security precautions are appropriate (e.g., which communication links to secure, databases requiring protection).
2. **Security Administration**—Vulnerabilities emerge because many systems lack a properly structured security policy (security administration is notoriously lax in the case of control systems), equipment and system implementation guides, configuration management, training, and enforcement and compliance auditing.
3. **Architecture**—Many common practices negatively affect SCADA security. For example, while it is convenient to use SCADA capabilities for other purposes such as fire and security systems, these practices create single points of failure. Also, the connection of SCADA networks to other automation systems and business networks introduces multiple entry points for potential adversaries.
4. **Network** (including communication links)—Legacy systems' hardware and software have very limited security capabilities, and the vulnerabilities of contemporary systems (based on modern information technology) are publicized. Wireless and shared links are susceptible to eavesdropping and data manipulation.
5. **Platforms**—Many platform vulnerabilities exist, including default configurations retained, poor password practices, shared accounts, inadequate protection for hardware, and nonexistent security monitoring

> controls. In most cases, important security patches are not installed, often due to concern about negatively impacting system operation; in some cases technicians are contractually forbidden from updating systems by their vendor agreements.

The following incident helps to illustrate some of the risks associated with SCADA vulnerabilities. During the course of conducting a vulnerability assessment, a contractor stated that personnel from his company penetrated the information system of a utility within minutes. Contractor personnel drove to a remote substation and noticed a wireless network antenna. Without leaving their vehicle, they plugged in their wireless radios and connected to the network within five minutes. Within 20 minutes they had mapped the network, including SCADA equipment, and accessed the business network and data. This illustrates what a cybersecurity advisor from Sandia National Laboratories specializing in SCADA stated—that utilities are moving to wireless communication without understanding the added risks.

THE INCREASING RISK

According to GAO (2003), historically, security concerns about control systems (SCADA included) were related primarily to protecting against physical attack and misuse of refining and processing sites or distribution and holding facilities. However, more recently there has been a growing recognition that control systems are now vulnerable to cyberattacks from numerous sources, including hostile governments, terrorist groups, disgruntled employees, hackers and "hactivists," and other malicious intruders.

In addition to control system vulnerabilities mentioned earlier, several factors have contributed to the escalation of risk to control systems, including (1) the adoption of standardized technologies with known vulnerabilities, (2) the connectivity of control systems to other networks, (3) constraints on the implementation of existing security technologies and practices, (4) insecure remote connections, and (5) the widespread availability of technical information about control systems.

ADOPTION OF TECHNOLOGIES WITH KNOWN VULNERABILITIES

When a technology is not well known, not widely used, not understood or publicized, it is difficult to penetrate it and thus disable it. Historically, proprietary hardware, software, and network protocols made it difficult to understand how control systems operated—and therefore how to hack into them. Today, however, to reduce costs and improve performance, organizations have been transitioning from proprietary systems to less expensive, standardized technologies such as Microsoft's Windows and Unix-like operating systems and the common networking protocols used by the Internet. These widely used standardized technologies have commonly known vulnerabilities, and sophisticated and effective exploitation tools are widely available and relatively easy to use. As a consequence, both the number of people with the knowledge to wage attacks and the number of systems subject to attack have increased. Also, common communication protocols and the emerging use of Extensible Markup Language (commonly referred to as XML) can make it easier for a hacker to interpret the content of communications among the components of a control system.

Control systems are often connected to other networks—enterprises often integrate their control system with their enterprise networks. This increased connectivity has significant advantages, including providing decision makers with access to real-time information and allowing engineers to monitor and control the process control system from different points on the enterprise network. In addition, the enterprise networks are often connected to the networks of strategic partners and to the Internet. Further, control systems are increasingly using wide area networks and the Internet to transmit data to their remote or local stations and individual devices. This convergence of control networks with public and enterprise networks potentially exposes the control systems to additional security vulnerabilities. Unless appropriate security controls are deployed in the enterprise network and the control system network, breaches in enterprise security can affect the operation of controls system.

According to industry experts, the use of existing security technologies, as well as strong user authentication and patch management practices, are generally not implemented in control systems because control systems operate in real time, typically are not designed with cybersecurity in mind, and usually have limited processing capabilities.

Existing security technologies such as authorization, authentication, encryption, intrusion detection, and filtering of network traffic and communications require more bandwidth, processing power, and memory than control system components typically have. Because controller stations are generally designed to do specific tasks, they use low-cost, resource-constrained microprocessors. In fact, some devices in the electrical industry still use the Intel 8088 processor, introduced in 1978. Consequently, it is difficult to install existing security technologies without seriously degrading the performance of the control system.

Further, complex passwords and other strong password practices are not always used to prevent unauthorized access to control systems, in part because this could hinder a rapid response to safety procedures during an emergency. As a result, according to experts weak passwords that are easy to guess, shared, and infrequently change are reportedly common in control systems, including the use of default passwords or *even no password at all.*

In addition, although modern control systems are based on standard operating systems, they are typically customized to support control system applications. Consequently, vendor-provided software patches are generally either incompatible or cannot be implemented without compromising service shutting down "always-on" systems or affecting interdependent operations.

Potential vulnerabilities in control systems are exacerbated by insecure connections. Organizations often leave access links—such as dial-up modems to equipment and control information—open for remote diagnostics, maintenance, and examination of system status. Such links may not be protected with authentication of encryption, which increases the risk that hackers could us these insecure connections to break into remotely controlled systems. Also, control systems often use wireless communications systems, which are especially vulnerable to attack, or leased lines that pass through commercial telecommunications facilities. Without encryption to protect data as it flows through these insecure connections or authentication mechanisms to limit access, there is limited protection for the integrity of the information being transmitted.

Public information about infrastructures and control systems is available to potential hackers and intruders. The availability of this infrastructure and vulnerability data was demonstrated by a university graduate student, whose dissertation reportedly mapped every business and industrial sector in the American economy to the fiber-optic network that connects them—using material that was available publicly on the Internet, none of which was classified. Many of the electric utility officials who were interviewed for the

National Security Telecommunications Advisory Committee's Information Assurance Task Force's Electric Power Risk Assessment expressed concern over the amount of information about their infrastructure that is readily available to the public.

In the electric power industry, for example, open sources of information—such as product data and educational videotapes from engineering associations can be used to understand the basics of the electrical grid. Other publicly available information—including filings of the Federal Energy Regulatory Commission (FERC), industry publications, maps, and material available on the Internet—is sufficient to allow someone to identify the most heavily loaded transmission lines and the most critical substations in the power grid.

In addition, significant information on control systems is publicly available—including design and maintenance documents, technical standards for the interconnection of control systems and RTUs, and standards for communication among control devise—all of which could assist hackers in understanding the systems and how to attack them. Moreover, there are numerous former employees, vendor, support contractors, and other end users of the same equipment worldwide with inside knowledge of the operation of control systems.

CYBER THREATS TO CONTROL SYSTEMS

As components of control systems increasingly make critical decisions that were once made by humans, the potential effect of a cyber threat becomes more devastating. Based on interviews and discussions with representatives throughout the electric power industry, the Information Assurance Task Force of the National Security Telecommunications Advisory Committee concluded that an organization with sufficient resources, such as a foreign intelligence service or a well-supported terrorist group, could conduct a structured attack on the electric power grid electronically, with a high degree of anonymity and without having to set foot in the target nation.

In July 2002, National Infrastructure Protection Center (NIPC) reported that the potential for compound cyber and physical attacks, referred to as "swarming attacks," is an emerging threat to the U.S. critical infrastructure. As NIPC reports, the effects of a swarming attack include slowing or complicating the response to a physical attack. For instance, a cyberattack that disabled the water supply or the electrical system in conjunction with a physical attack could deny emergency services the necessary resources to manage the consequences—such as controlling fires, coordinating actions, and generating light.

Entities or individuals with malicious intent might take one or more of the following actions to successfully attack SCADA/control systems:

- disrupt the operation of control systems by delaying or blocking the flow of information through control networks, thereby denying availability of the networks to control system operations;
- make unauthorized changes to programmed instructions in PLCs, RTUs, or DCS controllers, change alarm thresholds, or issue unauthorized commends to control equipment, which could potentially result in damage to equipment (if tolerances are exceeded), premature shutdown of processes (such as prematurely shutting down transmission lines), or even disabling of control equipment;
- send false information to control system operators either to disguise unauthorized changes or to initiate inappropriate actions by system operators;
- modify the control system software, producing unpredictable results; and
- interfere with the operation of safety systems.

In addition, in control systems that cover a wide geographic area, the remote sites are often unstaffed and may not be physically monitored. If such remote systems are physically breached, the attackers could establish a cyber connection to the control network.

SECURING CONTROL SYSTEMS

Several challenges must be addressed to effectively secure control systems against cyber threats. These challenges include: (1) the limitations of current security technologies in securing control systems; (2) the perception that securing control systems may not be economically justifiable; and (3) the conflicting priorities within organizations regarding the security of control systems.

A significant challenge in effectively securing control systems is the lack of specialized security technologies for these systems. The computing resources in control systems that are needed to perform security functions tend to be quite limited, making it very difficult to use security technologies within control system networks without severely hindering performance.

Securing control systems may not be perceived as economically justifiable. Experts and industry representatives have indicated that organizations may be reluctant to spend more money to secure control systems. Hardening the security of control systems would require industries to expend more resources, including acquiring more personnel, providing training for personnel, and potentially prematurely replacing current systems that typically have a lifespan of about twenty years.

Finally, several experts and industry representatives indicated that the responsibility for securing control systems typically includes two separate groups: IT security personnel and control system engineers and operators. IT security personnel tend to focus on securing enterprise systems, while control system engineers and operators tend to be more concerned with the reliable performance of their control systems. Further, they indicate that, as a result, those two groups do not always fully understand each other's requirements and collaborate to implement secure control systems.

IMPROVING SCADA SECURITY

The President's Critical Infrastructure Protection Board and the Department of Energy (DOE) have developed the steps outlined next to help organizations improve the security of their SCADA networks. DOE (2001) points out that these steps are not meant to be prescriptive or all-inclusive. However, they do address essential actions to be taken to improve the protection of SCADA networks. The steps are divided into two categories: specific actions to improve implementation, and actions to establish essential underlying management processes and policies.

21 Steps to Increase SCADA Security (DOE 2001)

The following steps focus on specific actions to be taken to increase the security of SCADA networks:

1. **Identify all connections to SCADA networks.**
   Conduct a thorough risk analysis to assess the risk and necessity of each connection to the SCADA network. Develop a comprehensive understanding of all connections to the SCADA network, and how well those connections are protected. Identify and evaluate the following types of connections:

- Internal local area and wide area networks, including business networks
- The Internet
- Wireless network devices, including satellite uplinks
- Modem or dial-up connections
- Connections to business partners, vendors, or regulatory agencies

2. **Disconnect unnecessary connections to the SCADA network.**
To ensure the highest degree of security of SCADA systems, isolate the SCADA network from other network connections to as great a degree as possible. Any connection to another network introduces security risks, particularly if the connection creates a pathway from or to the Internet. Although direct connections with other networks may allow important information to be passed efficiently and conveniently, insecure connections are simply not worth the risk; isolation of the SCADA network must be a primary goal to provide needed protection. Strategies such as utilization of "demilitarized zones" (DMZs) and data warehousing can facilitate the secure transfer of data from the SCADA network to business networks. However, they must be designed and implemented properly to avoid introduction of additional risk through improper configuration.

3. **Evaluate and strengthen the security of any remaining connections to the SCADA networks.**
Conduct penetration testing or vulnerability analysis of any remaining connections to the SCADA network to evaluate the protection posture associated with these pathways. Use this information in conjunction with risk management processes to develop a robust protection strategy for any pathways to the SCADA network. Since the SCADA network is only as secure as its weakest connecting point, it is essential to implement firewalls, intrusion detection systems (IDSs), and other appropriate security measures at each point of entry. Configure firewall rules to prohibit access from and to the SCADA network, and be as specific as possible when permitting approved connections. For example, an Independent System Operator (ISO) should not be granted "blanket" network access simply because there is a need for a connection to certain components of the SCADA system. Strategically place IDSs at each entry point to alert security personnel of potential breaches of network security. Organization

management must understand and accept responsibility or risks associated with any connection to the SCADA network.

4. **Harden SCADA networks by removing or disabling unnecessary services.** SCADA control servers built on commercial or open-source operating systems can be exposed to attack default network services. To the greatest degree possible, remove or disable unused services and network demons to reduce the risk of direct attack. This is particularly important when SCADA networks are interconnected with other networks. Do not permit a service or feature on a SCADA network unless a thorough risk assessment of the consequences of allowing the service/feature shows that the benefits of the service/feature far outweigh the potential for vulnerability exploitation. Examples of services to remove from SCADA networks include automated meter reading/remote billing systems, e-mail services, and Internet access. An example of a feature to disable is remote maintenance. Numerous secure configurations such as the National Security Agency's series of security guides. Additionally, work closely with SCADA vendors to identify secure configurations and coordinate any and all changes to operational systems to ensure that removing or disabling services does not cause downtime, interruption of service, or loss of support.

5. **Do not rely on proprietary protocols to protect your system.** Some SCADA systems are unique, proprietary protocols for communications between field devices and servers. Often the security of SCADA systems is based solely on the secrecy of these protocols. Unfortunately, obscure protocols provide very little "real" security. Do not rely on proprietary protocols or factor default configuration setting to protect your system. Additionally, demand that vendors disclose any backdoors or vendor interfaces to your SCADA systems, and expect them to provide systems that are capable of being secured.

6. **Implement the security features provided by device and system vendors.** Older SCADA systems (most systems in use) have no security features whatsoever, SCADA system owners must insist that their system vendor implement security features in the form of product patches or upgrades. Some newer SCADA devices are shipped with basic security features, but these are usually disabled to ensure ease of installation.

Analyze each SCADA device to determine whether security features are present. Additionally, factory default security settings (such as in computer network firewalls) are often set to provide maximum usability, but minimal security. Set all security features to provide the maximum security only after a thorough risk assessment of the consequences of reducing the security level.

7. **Establish strong controls over any medium that is used as a backdoor into the SCADA network.**
   Where backdoors or vendor connections do exist in SCADA systems, strong authentication must be implemented to ensure secure communications. Modems, wireless, and wired networks used for communications and maintenance represent a significant vulnerability to the SCADA network and remote sites. Successful "war dialing" or "war driving" attacks could allow an attacker to bypass all of other controls and have direct access to the SCADA network or resources. To minimize the risk of such attacks, disable inbound access and replace it with some type of callback system.

8. **Implement internal and external intrusion detection systems and establish 24-hour-a-day incident monitoring.**
   To be able to effectively respond to cyberattacks, establish an intrusion detection strategy that includes alerting network administrators of malicious network activity originating from internal or external sources. Intrusion detection system monitoring is essential 24 hours a day; this capability can be easily set up through a pager. Additionally, incident response procedures must be in place to allow an effective response to any attack. To complement network monitoring, enable logging on all systems and audit system logs daily to detect suspicious activity as soon as possible.

9. **Perform technical audits of SCADA devices and networks, and any other connected networks, to identify security concerns.**
   Technical audits of SCADA devices and networks are critical to ongoing security effectiveness. Many commercial and open-sourced security tools are available that allow system administrators to conduct audits of their systems/networks to identify active services, patch level, and common vulnerabilities. The use of these tools will not solve systemic

problems, but will eliminate the "paths of least resistance" that an attacker could exploit. Analyze identified vulnerabilities to determine their significance, and take corrective actions as appropriate. Track corrective actions and analyze this information to identify trends. Additionally, retest systems after corrective actions have been taken to ensure that vulnerabilities were actually eliminated. Scan nonproduction environments actively to identify and address potential problems.

10. **Conduct physical security surveys and assess all remote sites connected to the SCADA network to evaluate their security.**
    Any location that has a connection to the SCADA network is a target, especially unmanned or unguarded remote sites. Conduct a physical security survey and inventory access points at each facility that has a connection to the SCADA system. Identify and assess any source of information including remote telephone/computer network/fiber optic cables that could be tapped; radio and microwave links that are exploitable computer terminals that could be accessed; and wireless local area network access points. Identify and eliminate single points of failure. The security of the site must be adequate to detect or prevent unauthorized access. Do not allow "live" network access points at remote, unguarded sites simply for convenience.

11. **Establish SCADA "Red Teams" to identify and evaluate possible attack scenarios.**
    Establish a "Red Team" to identify potential attack scenarios and evaluate potential system vulnerabilities. Use a variety of people who can provide insight into weaknesses of the overall network, SCADA system, physical systems, and security controls. People who work on the system every day have great insight into the vulnerabilities of your SCADA network and should be consulted when identifying potential attack scenarios and possible consequences. Also, ensure that the risk from a malicious insider is fully evaluated, given that this represents one of the greatest threats to an organization. Feed information resulting from the "Red Team" evaluation into risk management processes to assess the information and establish appropriate protection strategies.

The following steps focus on management actions to establish an effective cyber security program:

12. **Clearly define cybersecurity roles, responsibilities, and authorities for managers, system administrators, and users.**
    Organization personnel need to understand the specific expectations associated with protecting information technology resources through the definition of clear and logical roles and responsibilities. In addition, key personnel need to be given sufficient authority to carry out their assigned responsibilities. Too often, good cybersecurity left up to the initiative of the individual, which usually leads to inconsistent implementations and ineffective security. Establish a cybersecurity organizational structure that defines roles and responsibilities and clearly identifies how cybersecurity issues are escalated and who is notified in an emergency.

13. **Document network architecture and identify systems that serve critical functions or contain sensitive information that require additional levels of protection.**
    Develop and document robust information security architecture as part of a process to establish an effective protection strategy. It is essential that organizations design their network with security in mind and continue to have a strong understanding of their network architecture throughout its lifecycle. Of particular importance, an in-depth understanding of the functions that the systems perform and the sensitivity of the stored information is required. Without this understanding, risk cannot be properly assessed and protection strategies may not be sufficient. Documenting the information security architecture and its components is critical to understanding the overall protection strategy, and identifying single points of failure.

14. **Establish a rigorous, ongoing risk management process.**
    A thorough understanding of the risks to network computing resources from denial-of-service attacks and the vulnerability of sensitive information to compromise is essential to an effective cybersecurity program. Risk assessments from the technical basis of this understanding and are

critical to formulating effective strategies to mitigate vulnerabilities and preserve the integrity of computing resources. Initially, perform a baseline risk analysis based on current threat assessment to use for developing a network protection strategy. Due to rapidly changing technology and the emergence of new threats on a daily basis, an ongoing risk assessment process is also needed so that routine changes can be made to the protection strategy to ensure it remains effective. Fundamental to risk management is identification of residual risk with a network protection strategy in place and acceptance of that risk by management.

15. **Establish a network protection strategy based on the principle of defense-in-depth.**
    A fundamental principle that must be part of any network protection strategy is defense-in-depth. Defense-in-depth must be considered early in the design phase of the development process, and must be an integral consideration in all technical decision-making associated with the network. Utilize technical and administrative controls to mitigate threats from identified risks to as great a degree as possible at all levels of the network. Single points of failure must be avoided, and cybersecurity defense must be layered to limit and contain the impact of any security incidents. Additionally, each layer must be protected against other systems at the same layer. For example, to protect against the inside threat, restrict users to access only those resources necessary to perform their job functions.

16. **Clearly identify cybersecurity requirements.**
    Organizations and companies need structured security programs with mandated requirements to establish expectations and allow personnel to be held accountable. Formalized policies and procedures are typically used to establish and institutionalize a cybersecurity program. A formal program is essential of establishing a consistent, standards-based approach to cybersecurity through an organization and eliminates sole dependence on individual initiative. Policies and procedures also inform employees of their specific cybersecurity responsibilities and the consequences of failing to meet those responsibilities. They also provide guidance regarding actions to be taken during a cybersecurity incident and promote efficient and effective actions during a time of crisis. As part of identifying cybersecurity requirements, include user agreements and notification and warning ban-

ners. Establish requirements to minimize the threat from malicious insiders, including the need for conducting background checks and limiting network privileges to those absolutely necessary.

17. **Establish effective configuration management processes.**
    A fundamental management process needed to maintain a secure network is configuration management. Configuration management needs to cover both hardware configurations and software configurations. Changes to hardware or software can easily introduce vulnerabilities that undermine network security. Processes are required to evaluate and control any change to ensure that the network remains secure. Configuration management begins with well-tested and documented security baselines for your various systems.

18. **Conduct routine self-assessments.**
    Robust performance evaluation processes are needed to provide organizations with feedback on the effectiveness of cybersecurity policy and technical implementation. A sign of a mature organization is one that is able to self-identify issues, conduct root cause analyses, and implement effective corrective actions that address individual and systemic problems. Self-assessment processes that are normally part of an effective cybersecurity program include routine scanning for vulnerabilities, automated auditing of the network, and self-assessments of organizational and individual performance.

19. **Establish system backups and disaster recovery plans.**
    Establish a disaster recovery plan that allows for rapid recover from any emergency (including a cyberattack). System backups are an essential part of any plan and allow rapid reconstruction of the network. Routinely exercise disaster recovery plans to ensure that they work and that personnel are familiar with them. Make appropriate changes to disaster recovery plans based on lessons learned from exercises.

20. **Senior organizational leadership should establish expectations for cybersecurity performance and hold individuals accountable for their performance.**
    Effective cybersecurity performance requires commitment and leadership from senior managers in the organization. It is essential that senior

management establish an expectation for strong cybersecurity and communicate this to their subordinate managers throughout the organization. It is also essential that senior organizational leadership establish a structure for implementation of a cybersecurity program. This structure will promote consistent implementation and the ability to sustain a strong cybersecurity program. It is then important for individuals to be held accountable for their performance as it relates to cyber security. This includes managers, system administrators, technicians, and users/operators.

21. **Establish policies and conduct training to minimize the likelihood that organizational personnel will inadvertently disclose sensitive information regarding SCADA system design, operations, or security controls.** Release data related to the SCADA network only on a strict, need-to-know basis, and only to persons explicitly authorized to receive such information. "Social engineering," the gathering of information about a computer or computer network via questions to naïve users, is often the first step in a malicious attack on computer networks. The more information revealed about a computer or computer network, the more vulnerable the computer/network is. Never divulge data revealed to a SCADA network, including the names and contact information about the system operators/administrators, computer operating systems, and/or physical and logical locations of computers and network systems over telephones or to personnel unless they are explicitly authorized to receive such information. Any requests for information by unknown persons need to be sent to a central network security location for verification and fulfillment. People can be a weak link in an otherwise secure network. Conduct training and information awareness campaigns to ensure that personnel remain diligent in guarding sensitive network information, particularly their passwords.

REFERENCES AND RECOMMENDED READING

Associated Press (AP). 2009. "Goal: Disrupt." *The Virginian-Pilot*, Norfolk, VA, April 4, 2009.

Brown, A. S. 2008. *SCADA vs. The Hackers*. American Society of Mechanical Engineers. http://www.memagazine.org/backissues/dec02/features/scadavs/ (accessed May 10, 2008).

DOE. 2001. *21 Steps to Improve Cyber Security of SCADA Networks*. Washington, DC: Department of Energy.

Ezell, B. C. 1998. *Risks of Cyber Attack to Supervisory Control and Data Acquisition*. VA: University of Virginia.

FBI. 2000. *Threat to Critical Infrastructure*. Washington, DC: Federal Bureau of Investigation.

FBI. 2004. *Ninth Annual Computer Crime and Security Survey*. Washington, DC: Computer Crime Institute and Federal Bureau of Investigations.

GAO. 2003. *Critical Infrastructure Protection: Challenges in Securing Control System*. Washington, DC: United States General Accounting Office.

Gellman, B. 2002. "Cyber-Attacks by Al Qaeda Feared: Terrorists at Threshold of Using Internet as Tool of Bloodshed, Experts Say." *Washington Post*, June 27, 2002, p. A01.

NIPC. 2002. *National Infrastructure Protection Center Report*. Washington, DC: National Infrastructure Protection Center.

Stamp, J., et al. 2003. *Common Vulnerabilities in Critical Infrastructure Control Systems*, 2nd ed. Sandia National Laboratories.

Stocks, C. 2014. "Water Power and Dam Construction." *Water Power Magazine*, 23 April.

USEPA. 2005. "EPA Needs to Determine What Barriers Prevent Water Systems from Securing Known SCADA Vulnerabilities." In J. Harris, *Final Briefing Report*. Washington, DC: USEPA.

Warwalking. 2003. http://warwalking.tribe.net (accessed May 9, 2008).

Young, M. A. 2004. *SCADA Systems Security*. SANS Institute.

8

Emergency Action

We're in uncharted territory.

—Rudi Giuliani (9/11/01)

Terrorism [takes] us back to ages we thought were long gone if we allow it a free hand to corrupt democratic societies and destroy the basic rules of international life.

—Jacques Chirac (1986)

To secure ourselves against defeat lies in our own hands, but the opportunity of defeating the enemy is provided by the enemy himself.

—Sun Tzu, *The Art of War*

When New York Mayor Rudi Giuliani made the earlier statement to Police Commissioner Bernard Kerik at the World Trade Center Site, September 11, 2001, to a point and to a degree, one of the first (and not to be forgotten) gross understatements of the twenty-first century had been uttered. Indeed, for citizens of the United States of America, the 9/11 events placed our level of consciousness, awareness, fear, and questions of what to do next in "uncharted territory." Actually, when you get right down to it, 9/11 generated more questions than anything else. Many are still asking the following questions today:

Why?

Why would anyone have the audacity to attack the United States?

What kind of cold-blooded killers would even think of conducting such an event?

Who were the Islamic radicals who perpetrated 9/11?

What were the terrorists' goal(s)?

Why?

Why were we not ready for such an attack?

Why had we not foreseen such an event?

Why were our emergency responders so undermanned, ill-prepared, and ill-equipped to handle such a disaster?

What took the military fighter planes so long to respond?

What did our government really know (if anything) before the events occurred?

Could anyone have prevented it?

Bottom line questions: Why us? Hell, why anyone?

Why?

These and several other questions continue to resonate today; no doubt they will continue to haunt us for some time to come.

Maybe we ask post 9/11-related questions because of who we are, what we are, and what we are not. That is, because we are Americans, we are free, uninhibited thinkers who think what we say and say what we think—isn't America great? Most Americans are soft-hearted and sympathetic to those in need—compassion is the very nature and soul of being American. Americans are not born terrorists; they are not born into a terrorist regime; they are not raised with fear in their hearts—they are not afraid every time they leave their homes and go about their daily business. Suicide bombers and other like terrorists are those that occupy some other faraway place; definitely not America, and they are definitely not American. Right?

Notwithstanding exceptions to the rule, such as Timothy McVeigh (a so-called red-blooded American, born and raised in America) and that other idiot (whether a national or foreigner) who mailed the anthrax in autumn 2001, terrorism was foreign to us.

Today, from a safety/security point of view, based on the events of 9/11 and the anthrax events, we should no longer be asking why. Instead, we should not waste our time, money, and energy asking why or in pointing a finger of blame at our government, military, 9/11 emergency responders, and/or the terrorists. We should stop asking *why* and shift our mindset to asking *what if*. The point is we need to stop feeling sorry for ourselves and accept the fact that there are folks out there who do not share our view of the American way of life. In chapter 6, in regards to security preparedness, we pointed to the need to ask what-if questions. Simply, what-if analysis is a proactive approach used to prevent or mitigate certain disasters, extreme events—whether human- or nature-generated. Obviously, asking and properly answering what-if questions has little effect on preventing the actions of Mother Nature, such as earthquakes, tornadoes, hurricanes (like Katrina), and others. On the other hand, it is true that what-if questions, when properly posed and answered (with results), can reduce the death toll and overall damage caused by these natural disasters. We are certainly aware that these natural events are possible, probable, and likely, and their effects can be horrendous—beyond tragic. The irony is apparent, however, especially when we ask: how many of us are actually willing to move away from or out of earthquake zones, tornado alleys, or floodplains to live somewhere else?

The fact is we do not possess a crystal ball to foretell the future. What-if questions prepare us to react and respond to certain contingencies. And respond we must, because there are certain events we simply can't prevent. The best response to an event we can't prevent is summed up by the Boy Scout motto: Be Prepared!

DAM SECTOR: WHAT-IF QUESTIONS

When it comes to the emergency action or response equation related to the Dam Sector, it is interesting to point out that there are two sides to this equation, and they are not necessarily balanced. For example, in a general emergency, on one side of the equation emergency response sometimes relies on Dam Sector assets for firefighting water supply, emergency water supply, and waterway access during a significant disaster where access by land is impossible. Again, there is another side to this equation: the potential crisis incidents side; this side of the equation is directly related to impact to the Dam Sector itself. Moreover, this side of the crisis incident equation might be better named the what-if side.

Why?

Well, it becomes clear when one asks the following what-if questions:

What if a Dam Sector asset is:

- attacked
- breached
- cyberattacked
- sabotaged
- a victim of earthquake
- a victim of equipment malfunction(s)
- a victim of excessive release
- a victim of facility misoperation
- a victim of lock closure
- a victim of physical security incident
- a victim of regulatory action
- a victim of a reservoir incident
- a victim of security posture modification
- a victim of significant inflow flood
- a victim of significant inflow of ice and debris
- a victim of structural modification
- a victim of suspicious activity
- a victim of vandalism/theft
- a victim of vessel allusion/collisions/grounding

ONE MIGHT ASK . . .

What if the Dam Sector is not prepared to deal with any of these what-if incidents? This is not an option; it is totally unacceptable. The Dam Sector and all other critical infrastructures understand the threats and the mitigation measures that are necessary to protect critical assets. Again, the Dam Sector recognizes the threats and has taken steps to respond to any and all incidents. The threats and planned mitigation steps are explained in the following sections.

DAM SECTOR THREATS

When planning for Dam Sector emergency response scenarios the full spectrum of intentional and unintentional threats must be considered. Such threats include:

- Natural threats (e.g., hurricane, fire, and floods)
- Manmade threats (e.g., chemical, radiological, and biological attacks)
- Workplace threats (e.g., pandemic flu, insider threat, and human error)
- Cyber-related threats (e.g., technological hazard and degradation of CAD system)

In addition to contagious human diseases and cyberattacks, complex, coordinated attacks with multiple events affect a large geographical region or multiple regions pose a significant threat to the Dam Sector. An attack of this magnitude would impact severely the sector's resilience and sustainability. Although the sector is robust and has built in redundant systems, multiple or widely distributed attacks could exhaust state, regional, and local resources.

DID YOU KNOW?

Threat capability is defined as the availability or the ease of use of tools or methods that could potentially be used to damage, disrupt, or destroy critical functions.

Unlike human-made threats, natural hazards are more predictable based on the availability of historical natural hazards data. Threat assessments are generally applicable to a geographic region or threat source. Figure 8.1 is a visual representation of different sources of threat.

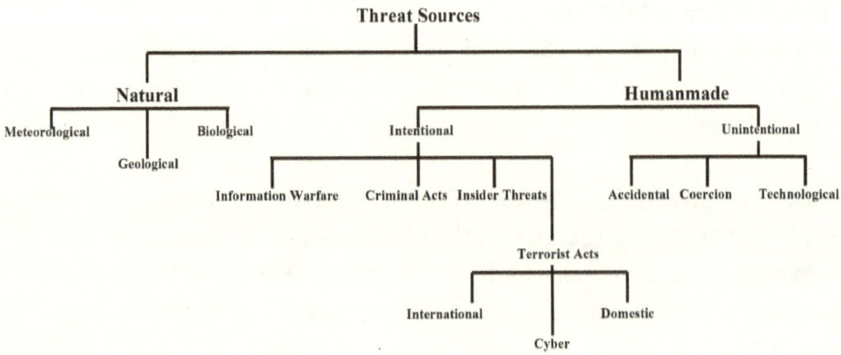

FIGURE 8.1
Threat sources.

DAM SECTOR CONTINGENCY PLANNING

Emergency response planning—or contingency planning or emergency action plans (EAPs)—for extreme events has long been standard practice for safety professionals in Dam Sector and other industrial systems operations. And, again, when we use the terms "Dam Sector," we are including the following components to perform their intended purposes safely: conventional dam or retaining structure (the structural sections that hold back the water), reservoir (the body of water impounded by the dam), spillways (the structure that passes normal and/or flood flows in a manner that protects the structural integrity of the dam), outlet works (the combination of structures and equipment required for safe operation and control of water released from a reservoir), powerhouse (the structure that houses turbines, generators, and associated control equipment for the production of hydroelectricity), and canal/aqueduct (a constructed channel, usually open, that conveys water by gravity or is used for navigation).

The bottom line: All Dam Sector systems are considered critical infrastructure.

For many years, prudent practices have required consideration of the potential impact of severe natural events (forces of nature), including earthquakes, tornadoes, volcanoes, floods, hurricanes, and blizzards. These possibilities have been included in Dam Sector infrastructure emergency preparedness and disaster response planning. In addition, many Dam Sector production and distribution facilities have considered the potential consequences of man-caused disasters such as operator error and manufacturer's industrial equipment defects. Currently, Dam Sector managers and operators must also consider violence in the workplace. Moreover, at present, as this text has pointed out, there is a new focus of concern: the potential effects of intentional acts by domestic (homegrown—with the emphasis on homegrown) or international (foreign) terrorists.

As a result, the security paradigm has not necessarily changed, but instead has been radically adjusted—reasonable, necessary, and sensible accommodation for and mitigation of just about any emergency situation imaginable have been and continue to be made. Because we cannot foresee all future domestic or intentional acts of terrorism, we must be prepared to shift from the proactive to reactive mode on short notice—in some cases, on very short notice. Accordingly, we must be prepared to respond to, react to, and mitigate what we can't prevent. Unfortunately, there is more we can't prevent than we can prevent. In light of this, in this chapter we present in outline form a reactive procedure designed to deal with acts of terrorism (domestic or foreign). In addition, we also present and describe another reactive mitigation procedure, the template for a standard Dam Sector Emergency Response or Contingency Plan, dealings with natural and man-caused disasters, which could also be applied in response to acts of terrorism.

EMERGENCY RESPONSE PLAN: STANDARD TEMPLATE

Note: The following criteria have not been established as anything other than guidelines and are offered not as definitive or official regulations or procedures but rather as informed advice (based on more than 30 years of safety, industrial hygiene, emergency contingency planning, and security experience) insofar as to the subject matter specific to both public and private sectors.

The goals of an Emergency Response Plan (ERP) are to document and understand the steps needed to:

- Rapidly restore energy producing, refining, transporting, generating, transmitting, conserving, building, distributing, maintaining, and controlling systems and components after an emergency.
- Minimize Dam Sector production process equipment damage.

- Minimize impact and loss to customers.
- Minimize negative impacts on public health and employee safety.
- Minimize adverse effects on the environment.
- Provide emergency public information concerning customer service.
- Provide hazardous chemical information for first responders and other outside agencies.

Although we are concerned with the Dam Sector in this text, the USEPA developed *Large Water System Emergency Response Outline: Guidance to Assist Community Water Systems in Complying with the Public Health and Bioterrorism Preparedness and Response Act of 2002* (dated July 2003), with minor adjustments, can be applied as a template for the Dam Sector. The Federal Energy Regulatory Commission also has developed a similar template for use by licensees. The template provides guidance and recommendations to aid facilities in the preparation of emergency response plans under the PL 107-188. The dam owner must ensure that the ERP conforms to any applicable State or Federal requirements. The ERP template is provided below.

Note: ERPs do not necessarily need to be one document. They may consist of an overview document, individual Emergency Action Procedures, check lists, additions to existing operations manuals, appendices, etc. There may be separate, more detailed plans for specific incidents. There may be plans that do not include particularly sensitive information and those that do. Existing applicable documents should be referenced in the ERP (e.g., chemical Risk Management Program, contamination response).

Also note that the ERP should be reviewed annually and updated as needed.

Dam Sector ERP Template

NOT FOR PUBLIC DISTRIBUTION

Emergency Response Plan

______________________ Watershed
Any county, Any town, Any State

Structure: __________________________
National Inventory of Dams ID#: __________

OWNER: ________________
ISSUE DATE: ______________

I. Introduction

Safe and reliable operation is vital to every industrial operation. Emergency Response Planning (ERP) is an essential part of managing a Dam Sector plant, process, or entity. The introduction should identify the requirement to have a documented emergency response plan (EFP), the goal(s) of the plan (e.g., be able to quickly identify an emergency and initiate timely and effective response action, be able to quickly respond and repair damages to minimize system downtime), and how access to the plan is limited. Plans should be numbered for control. Recipients should sign and date a statement that includes (1) their ERP number, (2) their agreement not to reproduce the ERP, and (3) that they have read the ERP.

II. Emergency Planning Process

A. Planning Partnerships

The planning process should include those parties who will need to help the Dam Sector in an emergency situation (e.g., first responders, law enforcement, public health officials, nearby utilities, local emergency planning committees, testing labs). Partnerships should track from the Dam Sector operation up through local, state, regional, and federal agencies, as applicable and appropriate, and could also document, compliance with governmental requirements.

B. General Emergency Response Policies, Procedures, Actions, Documents

A short synopsis of the overall emergency management structure, how other industrial emergency response, contingency, and risk management plans fit into the ERP for Dam Sector emergencies, and applicable policies, procedures, actions plans, and reference documents should be cited. Polices should include interconnect agreements with adjacent communities and just how the ERP may affect them.

C. Scenarios

Use your Vulnerability Assessment (VA) findings to identify specific emergency action steps required for response, recovery, and remediation for applicable incident types. In Section V of this plan, specific emergency actions procedures addressing each of the incident types should be addressed.

III. Emergency Response Plan—Policies

A. System Specific Information

In an emergency, Dam Sector assets need to have basic information for system personnel and external parties such as law enforcement, emergency responders, repair contractors/vendors, the media, and others. The information needs to be clearly formatted and readily accessible so system staff can find and distribute it quickly to those who may be involved in responding to the emergency. Basic information that may be presented in the emergency response plan are the system's ID number, system name, system address or location, directions to the system, population served, number of service connections, system owner, and information about the person in charge of managing the emergency. Distribution maps, detailed plant drawings, site plans, source/storage/production energy locations, and operations manuals may be attached to this plan as appendices or referenced.

1. PWS ID, Owner, Contact Person
2. Population served and service connections
3. System Components
   a) Pipes and constructed conveyances
   b) Physical barriers

c) Isolation valves
d) Chemical treatment, storage, and distribution facilities
e) Electronic, computer, or other automated systems which are utilized by the Dam Sector asset
f) Emergency power generators (onsite and portable)
g) The use, storage, or handling of various chemicals
h) The operation and maintenance of such system components

B. Chain-of-Command Chart (Notification Flowcharts) developed in Coordination with Local Emergency Planning Committee (internal and/or External Emergency Responders, or both)
1. Contact Name
2. Organization and Emergency Response Responsibility
3. Telephone number(s) (hardwire, cell phones, faxes, e-mail)
4. State 24-hour Emergency Communications Center Telephone

C. Communications Procedures: Who, What, When
During most emergencies, it will be necessary to quickly notify a variety of parties both internal and external to the Dam Sector entity. Using the Chain-of-Command Chart (Notification Flowcharts) and all appropriate personnel from the lists below, indicate who activates the plan, the order in which notification occurs, and the members of the Emergency Response Team. All contact information should be available for routine updating and readily available. The following lists are not intended to be all inclusive—they should be adapted to your specific needs.
1. Internal Notification Lists
a) Operations Dispatch
b) Dam Sector Plant Manager
c) Dam Sector Processing Manager
d) Dam Production/Storage/Distribution Manager
e) Facility Managers
f) Chief Energy Engineer
g) Director of Engineering
h) Data (IT) Manager
i) Maintenance Manager
j) Other

2. Local Notification
   a) Head of local government (i.e., Mayor, City Manager, Chairman of Board)
   b) Public Safety Officials—Fire, Local Law Enforcement (LLE), Police, EMS, Safety if a malevolent act is suspected, LLE should be immediately notified and in turn will notify the FBI, if required. The FBI is the primary agency for investigating sabotage to Dam Sector assets or terrorist incidents.
   c) Other Government Entities: Health, Schools, Parks, Finance, Electric, etc.
3. External Notification Lists
   a) State Department of Environmental Quality (DEQ)
   b) USEPA/USDOE/DHS
   c) State Police
   d) State Health Department (lab)
   e) Critical customers (special considerations for hospitals, Federal, State and Country government centers, etc.).
   f) Service Aid
   g) Mutual Aid
   h) Chemical Information Sharing and Analysis Center (ISAC)/CHEMTREC
   i) Commercial customers not previously notified
4. Public/Media Notification: When and How to Communicate Effective communications is a key element of emergency response, and a media or communications plan is essential to good communications. Be prepared by organizing basic facts about the crisis and your Dam Sector assets. Develop key messages to use with the media that are clear, brief, and accurate. Make sure your messages are carefully planned and have been coordinated with local and state officials. Considerations should be given to establishing protocols for both field and office staff to respectfully defer questions to the utility spokesperson.

   Be prepared to list geographic boundaries of the affected area (e.g., west of highway a, east of highway b, north of highway c and south of highway d to ensure the public clearly understand the system boundaries).

E. Personnel Safety

This should provide direction as to how operations staff, emergency responders, and the public should respond to a potential toxic release (e.g., electrical current released from downed transformers and lines), including facility evacuation, personnel accountability, proper Personnel Protective Equipment as dictated by the Risk Management Program and Process Safety Management Plan, and whether the nearby public should be "in-place sheltered" or evacuated.

F. Equipment

The ERP should identify equipment that can obviate or significantly lessen the impact of terrorist attacks or other intentional actions on the public health and protect the safety and supply of communities and individuals. The Dam Sector asset should maintain an updated inventory of current equipment and repair parts for normal maintenance work.

Because of the potential for extensive or catastrophic damage that could result from a malevolent act, additional equipment sources should be identified for the acquisition and installation of equipment and repair parts in excess of normal usage. This should be based on the results of the specific scenarios and critical assets identified in the vulnerability assessment that could be destroyed. For example, numerous pumps, vats, and mixers, specifically designed for the Dam Sector assets, could potentially be destroyed. A certain number of "long-lead" procurement equipment should be inventoried and the vendor information for such unique and critical equipment maintained. In addition, mutual aid agreements with other industries, and the equipment available under the agreement, should be addressed. Inventories of current equipment, repair parts, and associated vendors should be indicated under item CC. "Equipment Needs/Maintenance of Equipment" of Section IV "Emergency Action Procedures."

G. Property Protection

A determination should be made as to what Dam Sector producing, processing, distribution operation/facility should be immediately "locked down," specific access control procedures implemented, initial security perimeter established, a possible secondary malevolent event considered. The initial act may be a divisionary act.

H. Training, Exercises, and Drills

Emergency response training is essential. The purpose of the training program is to inform employees or what is expected of them during an emergency situation. The level of training on an ERP directly affects how well a Dam Sector facility's employees can respond to an emergency. This may take the form of orientation scenarios, table-top workshops, functional exercises, and so on.

I. Assessment

To evaluate the overall ERP's effectiveness and to ensure that procedures and practices developed under the ERP are adequate and are being implemented, the Dam Sector asset staff should audit the program on a periodic basis.

IV. Emergency Action Procedures

These are detailed procedures used in the event of an operation emergency of malevolent act. EAPs may be applicable across many different emergencies and are typically common core elements of the overall municipality ERP (e.g., responsibilities, notifications lists, security procedures) and can be referenced.

A. Event classification/severity of emergency
B. Responsibilities of Emergency Director
C. Responsibilities of incident Commander
D. Emergency Operations Center (EOC) activation
E. Division internal communications and reporting
F. External communications and notifications
G. Emergency telephone list (division internal contacts)
H. Emergency telephone list (off-site responders, agencies, state 24-hour emergency phone number, and others to be notified)
I. Mutual Aid Agreements
J. Contact list of available emergency contractor services/equipment
K. Emergency equipment list (including inventory for each facility)
L. Security and access control during emergencies
M. Facility evacuation and lockdown and personnel accountability
N. Treatment and transport of injured personnel (including electrocution and petrochemical exposure)

- O. Chemical records—to compare against historical results for base line
- P. List of available labs for emergency use
- Q. Emergency sampling and analysis (chemical)
- R. Water use restrictions during emergencies
- S. Alternate temporary chemical supplies during emergencies
- T. Isolation plans for chemical supply, treatment, storage, and distribution systems
- U. Mitigation plans for neutralizing, flushing, and collecting spilled chemicals
- V. Protection of vital records during emergencies
- W. Record keeping and reporting (FEMA, DHS, DOT, OSHA, EPA, and other requirements) (it is important to maintain accurate financial records of expenses associated with the emergency event for possible federal reimbursement)
- X. Emergency program training, drills/and tabletop exercises
- Y. Assessment of emergency management plan and procedures
- Z. Crime scene preservation training and plans
- AA. Communication Plans:
  1. Police
  2. Fire
  3. Local Government
  4. Media
  5. Etc.
- BB. Administration and logistics, including EOC, when established
- CC. Equipment needs/maintenance of equipment
- DD. Recovery and restoration of operations
- EE. Emergency event closeout and recovery

V. Incident-Specific Emergency Action Procedures

Incident-Specific EAPs are action procedures that identify specific steps in responding to an operational emergency of malevolent act.

- A. General Response to Terrorist Threats (Other than Bomb Threat and Incident-Specific Threats)
- B. Incident-Specific Response to Man-Made or Technological Emergencies
  1. Contamination event (Articulated Threat with Unspecified Materials)

    2. Contamination Threat at a Major Event
    3. Notification from Health Officials of Potential Contamination
    4. Intrusion through Supervisory Control and Data Acquisition (SCADA)

C. Significant structural damage resulting from intentional act
D. Customer complaints
E. Severe weather response (snow, ice, temperature, lightning)
F. Flood response
G. Hurricane and/or tornado response
H. Fire response
I. Explosion response
J. Major vehicle accident response
K. Electrical power outage response
L. Water supply interruption response
M. Transportation accident response—barge, plane, train, semi-trailer/tanker
N. Contaminated/tampered with water treatment chemicals
O. Earthquakes response
P. Disgruntled employees response (i.e., workplace violence)
Q. Vandals response
R. Bomb threat response
S. Civil disturbance/riot/strike
T. Armed intruder response
U. Suspicious mail handling and reporting
V. Hazardous chemical spill release response (including Material Safety Data Sheets)
W. Cybersecurity/SCADA system attack response (other than incident specific, e.g., hacker)

VI. Preparedness

Preparedness actions are taken to prevent a dam failure incident or to help reduce the effects of a dam failure and facilitate response to emergencies. The following actions describe some of the steps that could be taken at the dam to prevent or delay failure after an emergency is first discovered. **These actions should only be performed under the direction of the dam safety office or other qualified professional engineers.**

ACTIONS TO BE TAKEN IN THE EVENT OF:

Overtopping by Flood Waters:

- Provide erosion-resistant protection to the downstream slope by placing plastic sheets or other materials over eroding areas.
- Divert floodwaters around the reservoir basin, if possible.

A Slide on the Upstream or Downstream Slope of the Embankment:

- Lower the water level in the reservoir at a rate, and to an elevation, that is considered safe given the slide condition. If the outlet is damaged or blocked, pumping, siphoning, or a controlled breach may be required.
- Stabilize slides on the downstream slope by weighing the toe area below the slide with additional soil, rock, or gravel.

Erosional Seepage or Leakage (Piping) through the Embankment, Foundation, or Abutments:

- Plug the flow with whatever material is available (hay bales, bentonite, or plastic sheeting, if the entrance to the leak is in the reservoir).
- Lower the water level in the reservoir until the flow decreases to a non-erosive velocity or until it stops.
- Place an inverted filter (a protective sand and gravel filter) over the exit area to hold materials in place.
- Continue lowering the water level until a safe elevation is reached; continue operating at a reduced level until repairs are made.

Failure of an Appurtenant Structure Such as an Inlet/Outlet of Spillway:

- Implement temporary measures to protect the damaged structure, such as closing the inlet or providing temporary protection for a damaged spillway.
- Employ experienced, professional divers, if necessary, to assess the problem and possibly implement repair.
- Lower the water level in the reservoir to a safe elevation. If the inlet is inoperable, pumping, siphoning, or a controlled breach may be required.

Mass Movement of the Dam on Its Foundation (Spreading or Mass Sliding Failure):

- Immediately lower the water level until excessive movement stops.
- Continue lowering the water level until a safe level is reached; continue operation at reduced level until repairs are made.

Spillway Erosion Threatening Reservoir Evacuation:

- Provide temporary protection at the point of erosion by placing sandbags, riprap materials, or plastic sheets weighted with sandbags. Consider pumps and siphons to help reduce the water level in the reservoir.
- When inflow subsides, lower the water level in the reservoir to a safe level; continue operating at a lower water level in order to minimize spillway flow.

Excessive Settlement of the Embankment:

- Lower the water level by releasing it through the outlet or by pumping or siphoning.
- If necessary, restore freeboard, preferably by placing sandbags.
- Lower water level in the reservoir to a safe level; continue operating at a reduced level until repairs can be made.

Malicious Human Activity (Sabotage, Vandalism, or Terrorism)

- If malicious human activity that could endanger public safety is suspected, contact law enforcement to help evaluate the situation.
- If the principal spillway has been damaged or plugged, implement temporary measures to protect the damaged structure. Employ experienced, professional divers, if necessary, to assess the problem and possibly implement repair.
- If the embankment or spillway has been damaged or partially removed, provide temporary protection in the damaged area by placing sandbags, riprap materials, or plastic sheets weighed with sandbags. Use pumps and siphons to help reduce the water level in the reservoir.
- If the water supply has been contaminated, immediately close all inlets to the water supply system and notify appropriate authorities.

VII. Next Steps

A. Plan Review and Approval

B. Practice and Plan to Update (as necessary, once every year recommended)

1. Training requirements
2. Who is responsible for conducting training, exercises, and emergency drills?

3. Update and assessment requirements
4. Incident-specific requirements

VIII. Annexes
A. Facility and Location Information
1. Facility maps
2. Facility drawings
3. Facility descriptions/layout
4. Etc.

IX. References and Links
A. Department of Homeland Security—www.dhs/gov
B. Environmental Protection Agency—www.epa.gov
C. TheAmericanWaterWorksAssociation(AWWA)—www.awwa.org
D. Centers for Disease Control and Prevention—www. cdc.gov
E. Federal Emergency management Agency—www.fema.gov
F. Local Emergency Planning Committees [insert links as appropriate]

OSHA AND DAM SECTOR EMERGENCY RESPONSE PLANS

Even though no single OSHA standard is dedicated specifically to the issue of planning for Dam Sector facility/site/asset emergencies, all OSHA standards are written for the purpose of promoting a safe, healthy, accident-free, and hence emergency-free workplace. Therefore, OSHA standards do play a role in emergency prevention.

OSHA's standards, therefore, should be considered when developing emergency plans. A first step when developing ERPs is to review the guidelines presented earlier and then to add any and all applicable OSHA standards. This review of applicable guidelines and pertinent OSHA standards can help the managers of Dam Sector facilities identify and then correct conditions that might exacerbate emergency situations before they occur.

Typically, when we think of emergency response plans for the workplace, we often conjure up thoughts about the obvious. For example, the first workplace emergency that might come to mind is fire—a major concern because fire in the workplace is something that can happen, that happens more often than we might think, and because fire can be particularly devastating—in ways we know all too well. Most employees do not need to be informed about the dangers of fire. However, employers have the responsibility to do just this—to inform and train employees on fire, fire prevention, and fire protection. Many local codes go beyond this information requirement, insisting that employers develop and implement a fire emergency response and/or evacuation plan. The primary emphasis has been on the latter—evacuation. However, if the employer equips a workplace with fire extinguishers and other fire-fighting equipment, and expects its employees to respond aggressively to extinguish workplace fires, then not only must the facility have an ERP, the employer must also ensure that all company personnel called upon to fight the fire are completely trained on how to do so safely; this is stipulated in OSHA's 29 CFR 1910.156(c).157(g).

Another commonly considered workplace ERP or scenario is designed and implemented for medical emergencies. Many Dam Sector facilities satisfy this requirement simply by directing their employees to call 911 or some other emergency number whenever a medical emergency occurs in the workplace. Other facilities, though, may require employees to provide emergency first aid. When the employer chooses the employee-supplied first aid option, certain requirements must be met before any employee can legally administer first aid. First, the first aid responder must be trained and certified to administer first aid. This training aspect must also include training on OSHA's Bloodborne Pathogen Standard. This standard requires that the employee is trained on the dangers inherent with handling and being exposed to human body fluids. The employee must also be trained on how to protect him or herself from body fluid contamination. If the first aid responder or anyone else is exposed to and contaminated by body fluids, the employer must make available the hepatitis B vaccine and vaccination series to all employees who have occupational exposure, and post-exposure evaluation and follow-up to all employees who have had an exposure incident; this requirement is stipulated in OSHA's 29 CFR 1910.1030.

A third type of ERP required for implementation in selected (covered) facilities is OSHA's 29 CFR 1910.120 (Hazardous Waste Operations and Emergency Response—HAZWOPER)—for releases of hazardous petrochemical materials. Unless the facility operator can demonstrate that the operation does not involve employee exposure or the reasonable possibility for employee exposure to safety or health hazards, the following operations are covered:

1. Clean-up operations required by a governmental body involving hazardous substances conducted at uncontrolled hazardous waste sites, state priority site lists, sites recommended by the EPA, NPL, and initial investigations of government identified sites that are conducted before the presence or absence of hazardous substance has been ascertained.
2. Corrective actions involving clean-up operations at sites covered by the Resource Conservation and Recovery Act of 1976 (RCRA).
3. Voluntary clean-up operations at sites recognized by federal, state, local or other governmental bodies as uncontrolled hazardous waste sites.
4. Operations involving hazardous petrochemical waste conducted at treatment, storage, disposal (TSD) facilities regulated by RCRA.
5. Emergency response operations for releases of, or substantial threats of releases of, hazardous petrochemical substances without regard to the location of the hazard.

The final requirement impacts the largest number of facilities that meet the criteria requiring full compliance with 29 CFR 1910.120 HAZWOPER, because many such facilities do not normally handle, store, treat, or dispose of hazardous petrochemical waste, but do use or produce hazardous materials in their processes.

Because the use of hazardous petrochemical materials could lead to an emergency from the release or spill of such materials, facilities using these materials must develop and employ an effective site ERP.

Before we discuss the basic goals of an effective ERP from an OSHA compliance point of view, we should define "emergency response." Considering that individual facilities are different, with different dangers and different needs, defining emergency response is not always easy. However, for our purposes, we use the definition provided by CoVan (1995, p. 54).

> Emergency response is defined as a limited response to abnormal conditions expected to result in unacceptable risk requiring rapid corrective action to prevent harm to personnel, property, or system function.

CoVan makes another important point about emergency response, one critical for the site manager and/or safety professional. He points out that

DID YOU KNOW?

OSHA's 29 CFR 1910.151 (a), (b), and (c), Medical Services and First Aid standard, requires employers to (1) ensure the ready availability of medical personnel for advice and consultation on matters of health; (2) in the absence of an infirmary, clinic, or hospital in near proximity to the workplace which is used for the treatment of all injured employees, a person or persons shall be adequately trained to render first aid. Adequate first aid supplies shall be really available; and (3) where the eyes or body of any person may be exposed to injurious corrosive materials, suitable facilities for quick drenching of or flushing of the eyes and body shall be provided within the work area for immediate emergency use.

In auditing various industrial workplaces, we found that this particular OSHA standard is one of the most misunderstood by employers. Based on our experience, we found that OSHA's use of the words "near proximity to" and "adequate" contribute to the vagary and ambiguity of this standard. Most employers understand the need to provide first aid supplies in the workplace, but generally think that this is the extent of their responsibilities in this regard. Moreover, many workplaces do train their personnel on basic first aid and CPR, but do not require (in written job descriptions) that the trained employees respond to a workplace medical emergency.

We found that employers are reluctant to have employees respond to workplace medical emergencies because most employees do not want to respond. They do not want to be exposed to gore and the victim's pain. They also do not want to assume liability for trying to aid an injured victim.

The common response we received from workers can be summed up as follows:

> When it comes to liability matters in America, let me sue you before you sue me.

"although emergency response and engineering tends toward prevention, emergency response is a skill area that safety engineers must be familiar with both because of regulations and good engineering practice." "Good Engineering Practice" is the law by which all competent managers and safety professionals work and live.

Now that we have defined emergency response, let's move on to the basic goals of an effective ERP from an OSHA compliance viewpoint. The goals are usually defined as:

1. Minimize injury to facility personnel.
2. Minimize damage to facility and then return to normal operation as soon as possible.

Obviously, these goals make a great deal of good sense. However, you may be wondering about language used—a couple of key words: "facility personnel" and "damage to the facility." Remember that we are talking about OSHA requirements here. Under OSHA the primary emphasis is protecting the worker—protecting the worker's health and safety is OSHA's only focus.

What about people who live off site; the site's neighbors?

What about the environment?

These questions make the point we emphasize here. Again, OSHA is not normally concerned about the environment, unless contamination of the environment (at the worksite) might adversely impact the worker's safety and health. The neighbors? Again, OSHA's focus is the worker. One OSHA compliance officer explained to us that if the employer takes every necessary step to protect its employees from harm involving the use or production of hazardous chemical materials, then the surrounding community should have little to fear.

This statement is puzzling to us. We asked the same OSHA compliance officer about those incidents beyond the control of the employer—about accidents that could not only put employees in harm's way, but also endanger the surrounding community. The answer? "Well, that's the EPA's bag—we only worry about the worksite and the worker."

Fortunately, OSHA, in combination with the U.S. Environmental Protection Agency (USEPA), has taken steps to overcome this blatant shortcoming (we like to think of it as an oversight). Under OSHA's Process Safety

Management (PSM) and EPA's Risk Management Planning (RMP) directive, chemical spills and other chemical accidents that could impact both the environment and the "neighbors" have now been properly addressed. What PSM and RMP really accomplish is changing the typical twofold goal of an effective ERP to a threefold goal.

Let's point out that the accomplishment of these two- or three-fold goals or objectives is essential in emergency response. Accomplishing these goals or objectives requires an extensive planning effort prior to the emergency ("prior" being the keyword, because the attempts to develop an ERP when a disaster is occurring or after one has occurred is both futile and stupid). The site manager and/or safety professional must never forget that while hazards in any facility can be reduced, risk is an element of everyday existence, and therefore cannot be totally eliminated. The manager/safety professional's goal must be to keep risk to an absolute minimum. Again, to accomplish this, planning is critical—as well as essential, and should be accomplished well in advance of emergency, because program testing should be part of the overall program design.

It was pointed out earlier that most emergency plans address fire, medical emergencies, and the accidental release or spills of petrochemical materials. Note that the development of ERPs should also factor in other possible emergencies—natural disasters, floods, explosions, and/or weather-related events that could occur, and certainly will occur. Now, emergency response to terrorist activity or threats must also be added to the list.

Site emergency response plans should include:

- Assessment of risk
- Chain of command for dealing with emergencies
- Assessment of resources
- Training
- Incident command procedures
- Site security
- Public relations

The Federal Emergency Management Agency (FEMA), the U.S. Army Corps of Engineers, and several other agencies, as well as numerous publications, provide guidance on how to develop a site ERP. Local agencies (such as fire departments, emergency planning commissions/agencies, HazMat teams, and Local Emergency Planning Committees [LEPCs]) also provide information on how to design a site plan.

In the site manager/safety professional's effort to incorporate and manage a facility ERP, and in the response itself, two elements mentioned earlier (security considerations and public relations) must be given special attention. If not handled correctly, the lack of effective security measures and/or improper public relations can turn an already disastrous incident into a mega-disaster.

In planning security considerations, provision should be made to have a well-trained security team limit site access to only those people and that equipment that will assist in coping with and resolving the emergency.

To a certain extent, the person who is ultimately responsible for giving the "go-ahead" to starting up in response to a perceived emergency is in a "damned if you do, damned if you don't" position. When Hurricane Isabel hit the East Coast in 2003 with 100-mile-per-hour-plus winds and torrential rains, and as the effects of the storm were felt hundreds of miles away from the actual storm center, we all recognized that only slight changes in conditions could have pushed Isabel out to sea—and all that reaction to the approaching emergency would have looked like overkill. But that's both the beauty and the frustration of emergency planning. By calling in the process early, you may run the risk of looking overly cautious, or, heaven forbid, wimpy. But if the storm does hit, no one is going to remember that you made the right call—you just did your job. In general, Dam Sector site managers are people who would rather err on the side of caution (hopefully) . . . in other words, safety. Once the storm hits, it's too late to do much good if you weren't already geared up to go.

THE BOTTOM LINE

Because Dam Sector emergencies (in less than extreme conditions) can seriously affect the surrounding community and environment, and because poor planning and/or panic can only make a bad situation worse, and can also lead to additional injury and death, your role as Dam Sector site manager or site safety professional in emergency response is doubly important. A crisis out of hand can easily devastate a community—and your organization is (or should be) an active member of your community. By ensuring less than effective emergency response, Dam Sector site managers endanger not only themselves and their organizations but also endanger their organization's community and standing as well.

REFERENCES AND RECOMMENDED READING

CoVan, J. 1995. *Safety Engineering*. New York: Wiley.

Office of the Federal Register. 1987. 29 CFR 1910.120. Washington, DC: Office of the Federal Register.

Safety and Health Requirements Manual, rev. ed. October 1987. EM 385-1-1. Washington, DC: U.S. Army Corps of Engineers.

Spellman, F. R. 1997. *A Guide to Compliance for Process Safety Management Planning (PSM/RMP)*. Lancaster, PA: Technomic Publishing Company.

USDOE. 2008. *Emergency Support Function #12—Energy Annex*. Washington, DC: U.S. Department of Energy.

USEPA. 2002. *Water Utility Response, Recovery and Remediation Guidance for Man-Made and/or Technological Emergencies*. Washington, DC: United States Environmental Protection Agency.

USEPA. 2003. *Large Water System Emergency Response Plan Outline: Guidance to Assist Community Water Systems in Complying with the Public Health Security and Bioterrorism Preparedness and Response Act of 2002*. EPA 810-F-03-007. Washington, DC: United States Environmental Protection Agency. www.epa.gov/safewater/security (accessed June 6).

9

Security Techniques and Hardware

There are but two powers in the world, the sword and the mind. In the long run the sword is always beaten by the mind.

—Napoleon Bonaparte

We do not create terrorism by fighting the terrorists. We invite terrorism by ignoring them.

—George W. Bush, 2005

Ideally, in a perfect world, all Dam Sector sites/facilities/assets would be secured in a layered fashion. Layered security (aka multi-barrier) systems are vital. Using the "protection-in-depth" principle requiring that an adversary defeat several protective barriers or security layers to accomplish its goal, Dam Sector industry infrastructure can be made more secure. "Protection-in-depth" is a term commonly used by the military to describe security measures that reinforce one another, masking the defense mechanisms from view of intruders and allowing the defender time to respond to intrusion or attack.

A prime example of the use of the multi-barrier approach to ensure security and safety is demonstrated by the practices of the bottled water industry. In the aftermath of 9/11 and the increased emphasis on Homeland Security, a shifted paradigm of national security and vulnerability awareness emerged. Recall that in the immediate aftermath of the 9/11 tragedies, emergency responders and others responded quickly and worked to exhaustion. In addition to the emergency responders, bottled water companies responded immediately by donating several million bottles of water to the crews at the crash sites in New York, at the Pentagon, and in Pennsylvania. International Bottled Water Association (IBWA 2004) reports that "within hours of the first attack, bottled water was delivered where it mattered most; to emergency personnel on the scene who required ample water to stay hydrated as they worked to rescue victims and clean up debris."

Bottled water companies continued to provide bottled water to responders and rescuers at the 9/11 sites throughout the post-event(s) process(es). These patriotic actions by the bottled water companies, however, beg the question: How do we ensure the safety and security of the bottled water provided to anyone? IBWA (2004, p. 3) has the answer: Using a multi-barrier approach, along with other principles, will enhance the safety and security of bottled water. IBWA (2004) describes its multi-barrier approach as follows:

> **A multi-barrier approach**—Bottled water products are produced utilizing a multi-barrier approach, from source to finished product, that helps prevent possible harmful contaminants (physical, chemical, or microbiological) from adulterating the finished product as well as storage, production, and transportation equipment. Measures in a multi-barrier approach may include source protection, source monitoring, reverse osmosis, distillation, filtration, ozonation, or ultraviolet (UV) light. Many of the steps in a multi-barrier system may be effective in safeguarding bottled water from microbiological and other contamination. Piping in and out of plants, as well as storage silos and water tankers are also protected and maintained through sanitation procedures. In addition, bottled water products are bottled in a controlled, sanitary environment to prevent contamination during the filling operation.

So with Dam Sector infrastructure security, let's consider a protection-in-depth strategy: using several forms of security techniques and/or devices against an intruder that thus does not rely on one single defensive mechanism to protect infrastructure. By implementing multiple layers of security, a hole or flaw in one layer is covered by the other layers. An intruder will have to intrude through each layer without being detected in the process—the layered approach implies that no matter how an intruder attempts to accomplish his or her goal, he or she will encounter effective elements of the physical protection system.

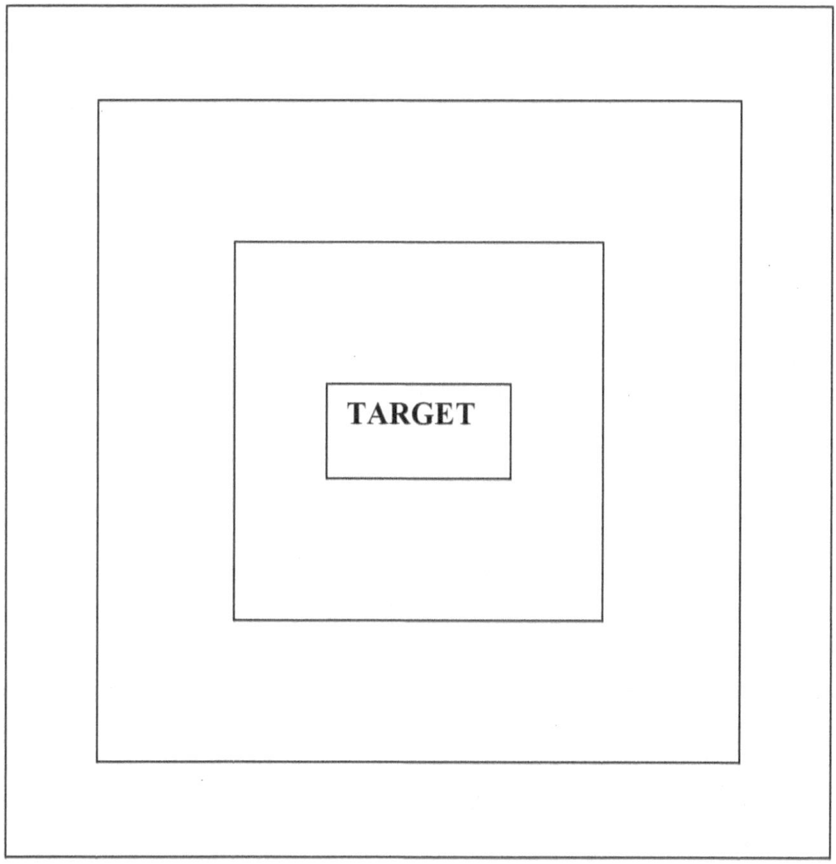

FIGURE 9.1
Multiple layers of security.

For example, look at figure 9.1: an effective security layering approach requires that an adversary penetrate multiple, separate barriers to gain entry to a critical TARGET at a dam facility. As shown in figure 9.1, protection in depth (multiple layers of security) helps to ensure that the security system remains effective in the event of a failure or an intruder bypassing a single layer of security.

In figure 9.1, layered security starts with the outer perimeter (the fence—the first line of physical security) of the facility and goes inward to the facility, the buildings, structures, other individual assets, and finally to the contents of those buildings—the TARGETs.

The area between the outer perimeter and structures or buildings is known as the site. This open site area provides an incomparable opportunity for early identification of an unauthorized intruder and initiation of early warning/ response. This open space area is commonly used to calculate the standoff distance; that is, it is the distance between the outside perimeter (public areas to the fence) to TARGET or critical assets (buildings/structures) inside the perimeter (inside the fence line—the restricted access area).

The open area, between perimeter fence and TARGET (e.g., operations center), if properly outfitted with various security devices, can also provide layered protection against intruders. For example, lighting is a deterrent. Based on personal experience, I think an open area within the facility site should be almost as well lighted at night as would be experienced during day. In addition, strategically placed motion detectors along with crash barriers at perimeter gate openings and in front of vital structures are also recommended. Of course, armed, mobile guards who roam the interior of the site on a regular basis would provide the ultimate in site area security.

The next layer of physical security is the outside wall of the target structure(s) itself. Notwithstanding door, window and/or skylight entry, walls prevent most intruders from easy entry. If doors can only be entered using card reader access, security is shored up or enhanced to an extent. The same can be said for windows and skylights that are fashioned small enough to prohibit normal human entry. These same "weak" spots in buildings can be bastioned with break-proof or reinforced security glass.

The final layer of security is provided by properly designed interior features of buildings. Examples of these types of features include internal doors and walls, equipment cages, and backup or redundant equipment.

In the preceding paragraphs, the layered security features described are in "perfect world" conditions; that is, those conditions that we "would want" (i.e., the security manager's proverbial wish list) to be incorporated into the design and installation of new Dam Sector infrastructure and assets. When it comes to providing security for Dam Sector assets there is a fly in the ointment; a huge fly. That is, certain characteristics of dams make them an unusually difficult type of asset to protect. While the critical assets in many other sectors are small or concentrated and can be contained within buildings, dams are very large, most components are not within buildings, dams are often in remote locations, and they can be approached via land, water, or even air. These factors pose especially difficult problems in controlling access to dams. In addition, certain federal agencies and FERC licensees are required to provide public access to certain portions of the dam project.

Post 9/11, in a not-so-perfect world and for the reasons mentioned above, many of the peripheral (fence line) measures described above are more difficult to incorporate into Dam Sector infrastructure and assets. This is not to say that Dam Sector sites and facilities do not have fence lines or fences; many of them do. These fences are designed to keep vandals, thieves, and trespassers out. The problem is not only the fact that many of these facilities were constructed years or decades before urban encroachment literally encircled the many sites, allowing, at present, little room for security stand backs or setbacks to be incorporated into electrical power stations, plants, and/or critical equipment locations. In my own tours of many facilities, I've observed that many of their fences face busy city streets or closely abut structures outside the fence line. My point is that when one sits down to plan a security upgrade, these factors must be taken into account.

The problem in protecting Dam Sector sites/facilities/equipment is much bigger, about 100 times more complicated, than urban encroachment, however. Actually, the *lack* of urban encroachment is one of the main problems in trying to secure Dam Sector sites. Think about it!

The electrical sector has many hydropower dams and more than 3,300 power plants and transmission lines that extend here, there, and everywhere for more than 211,000 miles across the contiguous United States. So, again, the problem with protecting the Dam Sector is one of distance, wide dispersion and remoteness—complicated by hackability of cyber networks and the SCADA grid control systems.

The multiple-barrier approach to security that was described earlier can be employed with relative ease to other critical infrastructures such as chemical sites, defense industrial bases, postal facilities, public health centers, monuments and icons, commercial assets, and other entities. The same cannot be said for protecting many of the remote Dam Sector assets and electrical transmission towers— many of which are 150 feet in height, with cross-arms as much as 100 feet wide, and thus the most visible component of electric transmission systems. To a limited degree, stepped-up security patrols is the partial answer to security concerns of the remote infrastructure such as Dam Sector assets and electrical transmission towers.

Managers of Dam Sector infrastructure have four primary security areas to manage:

- **Physical Security**—in the Dam Sector physical security techniques and practices have the most effective at sites such as electrical substations and other "fenced" locations. At such locations, a systems approach is best, where detection, assessment, communication, and response are planned and supported by resources, procedures, and policies.
- **Cyber/Information Technology Security**—only the use of SCADA and other key operating systems that have been properly vetted and scrubbed of alleged Chinese and/or Russian Trojan Horses hacked into the North American electrical grid is important. The only positive way to ensure the security of the North American Grid is to disconnect its cyber and other digital systems from the Internet. This step is impractical at the present time, but points to the need to conduct frequent audits of the system and install firewall protection in SCADA and other systems to prevent hacking. Frequent third-party penetration testing is advised.
- **Employment Screening**—mitigates the threat from the enemy at the water cooler (inside the organization). We are always amazed whenever we conduct security audits for various companies. Often, a simple check such as reviewing an employee's driving record often reveals that the employee has no license, is driving on a suspended license, or has a horrific driving record. Hiring standards and pre-employment background investigations may help ensure the trustworthiness and reliability of personnel who have unescorted access to critical facilities.
- **Protecting Potentially Sensitive Information**—the old saying that goes—a secret can best be held between three people so long as two of the three are dead—makes the point that reducing the likelihood that information could be used by those intent on disrupting operations or causing death destruction in Dam Sector plants/sites is crucial. Information should only be shared within an organization on a need-to-know basis only.

For existing facilities, security upgrades should be based on the results generated from the vulnerability assessment (chapter 5), which characterizes and prioritizes those assets that may be targeted. Those vulnerabilities identified must be protected.

In the following sections, various security hardware and/or devices are described. These devices serve the main purpose of providing security against physical and/or digital intrusion. That is, they are designed to delay and deny intrusion and are normally coupled with detection and assessment technology. Possible additional security measures, based on the vulnerability assessment that may be recommended (covered in this text) include the following (NAERC 2002):

- Electronic Security
- Closing nonessential perimeter and internal portals
- Physical barriers such as bollards or highway (Jersey) walls
- Fencing
- Lighting
- Security surveys
- Vulnerability Assessments
- Availability of security resources
- General personnel and security officer training
- Law Enforcement Liaison
- Ensure availability of essential spare parts (wire, pipe, valves, transformers, etc.) for critical facilities

Again, keep in mind, though, that no matter the type of security device or system you employ, Dam Sector systems cannot be made immune to all possible intrusions or attacks. Whenever a plant or facility safety/security manager tells you that he or she has secured their site 100 percent, we are reminded of Bruce Schneier's (2000) view of security: "You can't defend. You can't prevent. The only thing you can do is detect and respond [or pay your respects to the penitent ashes]." Simply, when it comes to making "anything" absolutely secure from intrusion or attack, there is inherently, or otherwise, no silver bullet.

SECURITY HARDWARE DEVICES[1]

The USEPA (2005) groups infrastructure security devices or products described below into four general categories:

- Physical asset monitoring and control devices
- Cyber protection devices
- Communication/integration
- Environmental monitoring devices

PHYSICAL ASSET MONITORING AND CONTROL DEVICES—ABOVEGROUND, OUTDOOR ENCLOSURES

As mentioned, Dam Sector facilities and sites usually consist of multiple components spread over a wide area (e.g., transmission towers and pipelines reaching thousands of acres and miles in some cases) and typically include a centralized distribution and control center, as well as ancillary components that are typically distributed at multiple locations throughout the area. Dam Sector site/plant system designers have favored placing critical equipment—especially assets that require regular use and maintenance—aboveground. One of the primary reasons for doing so is that locating this equipment aboveground eliminates the safety risks associated with confined space entry, which is often required for the maintenance of equipment located belowground. With regard to electrical power, electrical substations are not usually suited for underground installation. Therefore, many pieces of critical electrical equipment are located outdoors and aboveground in configurations that are properly fenced, insulated or isolated to prevent accidental electrical shock or short circuits/fires in equipment.

Experience demonstrates that many different system components can be and are often installed outdoors and aboveground. Examples of these types of components could include:

- Backflow prevention devices
- Air release and control valves
- Pressure vacuum breakers

1. It is important to point out that even though the following USEPA security asset and device recommendations were first made for the water/wastewater critical infrastructure; these recommendations are applicable to all other critical infrastructure sectors, including Dam Sector assets.

- Oil and gas pumps and motors
- Chemical storage and feed equipment
- Meters
- Sampling equipment
- Instrumentation
- Electrical substations

Much of this equipment is installed in remote locations and/or in areas where the public (and terrorists) can access it.

One of the most effective security measures for protecting aboveground equipment, where feasible, is to place it inside a building or exterior fenced structure. When/where this is not possible, enclosing the equipment or parts of the equipment using some sort of commercial or homemade add-on structure may help to prevent tampering with the equipment. These types of add-on structures or enclosures, which are designed to protect people and animals from electrocution and to protect equipment both from the elements and from unauthorized access or tampering, typically consist of a box-like fenced structure that is placed over or around the entire component, or over/around critical parts of the component (i.e., valves), and is then secured to delay or prevent intruders from tampering with the equipment. The enclosures are typically locked or otherwise anchored to a solid foundation, which makes it difficult for unauthorized personnel to remove the enclosure and access the equipment.

Standardized aboveground enclosures are available in a wide variety of materials, sizes, and configurations. Many options and security features are also available for each type of enclosure, and this allows system operators the flexibility to customize an enclosure for a specific application and/or price range. In addition, most manufacturers can custom-design enclosures if standard, off-the-shelf enclosures do not meet a user's needs.

Many of these enclosures are designed to meet certain standards. For example, the American Society of Sanitary Engineers (ASSE) has developed Standard #1060, *Performance Requirements for Outdoor Enclosures for Backflow Prevention Assemblies.* If an enclosure will be used to house backflow preventer, this standard specifies the acceptable construction materials for the enclosure, as well as the performance requirements that the enclosure should meet, including specifications for freeze protection, drainage, air inlets, access for maintenance, and hinge requirements. ASSE #1060 also states that the enclosure should be lockable to enhance security.

Electrical substation and electrical equipment enclosures must meet the requirements and recommendations of various OHSA Standards, National Fire Protection Association (NFPA), National Electrical Codes (NEC), Institute of Electrical and Electronic Engineers (IEEE), and local code requirements.

Equipment enclosures can generally be categorized into one of four main configurations, which include:

- One-piece, drop-over enclosures
- Hinged or removable top enclosures
- Sectional enclosures
- Shelters with access locks

All enclosures, including those with integral floors, must be secured to a **foundation** to prevent them from being moved or removed. Unanchored or poorly anchored enclosures may be wind-blown off the equipment being protected, or may be defeated by intruders. In either case, this may result in the equipment beneath the enclosure becoming exposed and damaged. Therefore, ensuring that the enclosure is securely anchored will increase the security of the protected equipment.

The three basic types of foundations that can be used to anchor the aboveground equipment enclosure are concrete footers, concrete slabs-on-grade, or manufactured fiberglass pads. The most common types of foundations utilized for equipment enclosures are standard or slab-on-grade footers; however, local climate and soil conditions may dictate whether either of these types of foundations can be used. These foundations can be either precast or poured in place at the installation site. Once the foundation is installed and properly cured, the equipment enclosure is bolted or anchored to the foundation to secure it in place.

An alternative foundation, specifically for use with smaller Hot Box® enclosures, is a manufactured fiberglass pad known as the Glass Pad™. The Glass Pad™ has the center cut out so that it can be dropped directly over the piece of equipment being enclosed. Once the pad is set level on the ground, it is backfilled over a two-inch flange located around its base. The enclosure is then placed on top of the foundation, and is locked in place with either a staple- or a slotted-anchor, depending on the enclosure configuration.

One of the primary attributes of a security enclosure is its strength and resistance to breaking and penetration. Accordingly, the **materials** from which the enclosure is constructed will be important in determining the strength of the enclosure, and thus its usefulness for security applications. Enclosures are typically manufactured from either fiberglass or aluminum. With the exception of the one piece, drop-over enclosure, which is typically fabricated from fiberglass, each configuration described above can be constructed from either material. In addition, enclosures can be custom-manufactured from polyurethane, galvanized steel, or stainless steel. Galvanized or stainless steel is often offered as an exterior layer, or "skin," for an aluminum enclosure. Although they are typically utilized in underground applications, precast concrete structures can also be used as aboveground equipment enclosures. However, precast structures are much heavier and more difficult to maneuver than are their fiberglass and aluminum counterparts. Concrete is also brittle, and that can be a security concern, however, products can be applied to concrete structures to add strength and minimize security risks (i.e., epoxy coating). Because precast concrete structures can be purchased from any concrete producers, this document does not identify specific vendors for these types of products.

In addition to the construction materials, **enclosure walls** can be configured or reinforced to give them added strength. Adding insulation is one option that can strengthen the structural characteristics of an enclosure; however, some manufacturers offer additional features to add strength to exterior walls. For example, while most enclosures are fabricated with a flat wall construction, some vendors manufacture fiberglass shelters with ribbed exterior walls. These ribs increase the structural integrity of the wall and allow the fabrication of standard shelters up to twenty feet in length. Another vendor has developed a proprietary process that uses a series of integrated fiberglass beams that are placed throughout a foam inner core to tie together the interior and exterior walls and roof. Yet another vendor constructs aluminum enclosures with horizontal and vertical redwood beams for structural support.

Other security features that can be implemented on aboveground, outdoor equipment enclosures include locks, mounting brackets, tamper-resistant doors, and exterior lighting.

PHYSICAL ASSET MONITORING AND CONTROL DEVICES—ACTIVE SECURITY BARRIERS

Active security barriers (also known as crash barriers) are large structures that are placed in roadways at entrance and exit points to control or prevent vehicle access to these areas. These barriers are placed perpendicular to traffic to block the roadway, so that the only way that traffic can pass the barrier is for the barrier to be moved out of the roadway. These types of barriers are typically constructed from sturdy materials, such as concrete or steel, so that vehicles cannot penetrate through them. They are also designed at a certain height off the roadway so that vehicles cannot go over them.

The key difference between *active* security barriers—which include wedges, crash beams, gates, retractable bollards, and portable barricades—and *passive* security barriers—which include non-moveable bollards, jersey barriers, and planters—is that active security barriers are designed so that they can be raised and lowered or moved out of the roadway easily to allow authorized vehicles to pass them. Many of these types of barriers are designed so that they can be opened and closed automatically (i.e., mechanized gates, hydraulic wedge barriers), while others are easy to open and close manually (swing crash beams, manual gates). In contrast to active barriers, passive barriers are permanent, non-movable barriers, and thus they are typically used to protect the perimeter of a protected facility, such as sidewalks and other areas that do not require vehicular traffic to pass them. Several of the major types of active security barriers such as wedge barriers, crash beams, gates, bollards, and portable/removable barricades are described next.

Types of Active Security Barriers

Wedge barriers are plated, rectangular steel buttresses approximately two to three feet high that can be raised and lowered from the roadway. When they are in the open position, they are flush with the roadway and vehicles can pass over them. However, when they are in the closed (armed) position, they project up from the road at a 45 degree angle, with the upper end pointing toward the oncoming vehicle and the base of the barrier away from the vehicle. Generally, wedge barriers are constructed from heavy gauge steel, or concrete that contains an impact-dampening iron rebar core that is strong and resistant to breaking or cracking, thereby allowing them to withstand the impact from a vehicle attempting to crash through them. In addition, both of these materials help to transfer the energy of the impact over the barrier's entire volume, thus helping to prevent the barrier from being sheared off its base. In addition, because the barrier is angled away from traffic, the force of any vehicle impacting the barrier is distributed over the entire surface of the barrier and is not concentrated at the base, which helps prevent the barrier from breaking off at the base. Finally, the angle of the barrier helps hang up any vehicles attempting to drive over it.

Wedge barriers can be fixed or portable. Fixed wedge barriers can be mounted on the surface of the roadway ("surface-mounted wedges") or in a shallow mount in the road's surface, or they can be installed completely below the road surface. Surface-mounted wedge barricades operate by rising from a flat position on the surface of the roadway, while shallow-mount wedge barriers rise from their resting position just below the road surface. In contrast, below-surface wedge barriers operate by rising from beneath the road surface. Both the shallow-mounted and surface-mounted barriers require little or no excavation, and thus do not interfere with buried utilities. All three barrier mounting types project above the road surface and block traffic when they are raised into the armed position. Once they are disarmed and lowered, they are flush with the road; thereby allowing traffic to pass portable wedge barriers that are moved into place on wheels that are removed after the barrier has been set into place.

Installing rising wedge barriers requires preparation of the road surface. Installing surface-mounted wedges does not require that the road be excavated; however, the road surface must be intact and strong enough to allow the bolts anchoring the wedge to the road surface to attach properly. Shallow-mount and below-surface wedge barricades require excavation of a pit that is large enough to accommodate the wedge structure, as well as any arming/disarming mechanisms. Generally, the bottom of the excavation pit is lined with gravel to allow for drainage. Areas not sheltered from rain or surface runoff can install a gravity drain or self-priming pump. Table 9.1 lists the pros and cons of wedge barriers.

Table 9.1. Pros and Cons of Wedge Barriers

Pros	*Cons*
Can be surface-mounted or completely installed below the roadway surface. Wedge barriers have a quick response time (normally 3.5–10.5 seconds, but barrier can be 1–3 seconds in emergency situations. Because emergency activation of the barrier causes more wear and tear on the system than does normal activation, it is recommended for use only in true emergency situations. Surface or shallow-mount wedge barricades can be utilized in locations with a high water table and/or corrosive soils. All three wedge barrier designs have a high crash rating, thereby allowing them to be employed for higher security applications. These types of barrier are extremely visible, which may deter potential intruders.	Installations below the surface of the roadway will require construction that may interfere with buried utilities. Regular maintenance is needed to keep wedge fully operational. Improper use of the system may result in authorized vehicles being hung up by the barrier and damaged. Guards must be trained to use the system properly to ensure that this does not happen. Safety technologies may also be installed to reduce the risk of the wedge activating under an authorized vehicle.

Source: USEPA, 2005.

Crash beam barriers consist of aluminum beams that can be opened or closed across the roadway. While there are several different crash beam designs, every crash beam system consists of an aluminum beam that is supported on each side of the made by a solid footing or buttress, which is typically constructed from concrete, steel, or some other strong material. Beams typically contain an interior steel cable (typically at least one inch in diameter) to give the beam added strength and rigidity. The beam is connected by a heavy duty hinge or other mechanism to one of the footings so that it can swing or rotate out of the roadway when it is open, and can swing back across the road when it is in the closed (armed) position, blocking the road and inhibiting access by unauthorized vehicles. The nonhinged end of the beam can be locked into its footing, thus providing anchoring for the beam on both sides of the road and increasing the beam's resistance to any vehicles attempting to penetrate through it. In addition, if the crash beam is hit by a vehicle, the aluminum beam transfers the impact energy to the interior cable, which in turn transfers the impact energy through the footings and into their foundation, thereby minimizing the chance that the impact will snap the beam and allow the intruding vehicle to pass through.

Crash beam barriers can employ drop-arm, cantilever, or swing beam designs. Drop-arm crash beams operate by raising and lowering the beam vertically across the road. Cantilever crash beams are projecting structures that are opened and closed by extending the beam from the hinge buttress to the receiving buttress located on the opposite side of the road. In the swing beam design, the beam is hinged to the buttress such that it swings horizontally across the road. Generally, swing beam and cantilever designs are used at locations where a vertical lift beam is impractical. For example, the swing beam or cantilever designs are utilized at entrances and exits with overhangs, trees, or buildings that would physically block the operation of the drop-arm beam design.

Installing any of these crash beam barriers involves the excavation of a pit approximately 48 inches deep for both the hinge and the receiver footings. Due to the depth of excavation, the site should be inspected for underground utilities before digging begins. Table 9.2 lists the pros and cons of crash beams.

Table 9.2. Pros and Cons of Crash Beams

Pros	*Cons*
Requires little maintenance, while providing long-term durability. No excavation is required in the roadway itself to install crash beams.	Crash beams have a slower response time (normally 9.5–15.3 seconds, but can be reduced to 7–10 seconds in emergency situations) than do other types of active security barriers, such as wedge barriers. Because emergency activation of the barrier causes more wear and tear on the system than does normal activation, it is recommended for use only in true emergency situations. All three crash beam designs possess a low crash rating relative to other types of barriers, such as wedge barriers, such as wedge barriers, and thus they typically are used for lower security applications. Certain crash barriers may not be visible to oncoming traffic and therefore may require additional lighting and/or other warning markings to reduce the potential for traffic to accidentally run into the beam.

Source: USEPA, 2005.

In contrast to wedge barriers and crash beams, which are typically installed separately from a fence line, **gates** are often integrated units of a perimeter fence or wall around a facility. Gates are basically movable pieces of fencing that can be opened and closed across a road. When the gate is in the closed (armed) position, the leaves of the gate lock into steel buttresses that are embedded in concrete foundation located on both sides of the roadway, thereby blocking access to the roadway. Generally, gate barricades are constructed from a combination of heavy gauge steel and aluminum that can absorb an impact from vehicles attempting to ram through them. Any remaining impact energy not absorbed by the gate material is transferred to the steel buttresses and their concrete foundation.

Gates can utilize a cantilever, linear, or swing design. Cantilever gates are projecting structures that operate by extending the gate from the hinge footing across the roadway to the receiver footing. A linear gate is designed to slide across the road on tracks via a rack and pinion drive mechanism. Swing gates are hinged so that they can swing horizontally across the road.

Installation of the cantilever, linear, or swing gate designs described earlier involve the excavation of a pit approximately 48 inches deep for both the hinge and receiver footings to which the gates are attached. Due to the depth of excavation, the site should be inspected for underground utilities before digging begins. Table 9.3 lists the pros and cons of gates.

Bollards are vertical barriers at least 3 feet tall and 1 to 2 feet in diameter that are typically set 4 to 5 feet apart from each other so that they block vehicles from passing between them. Bollards can either be fixed in place, removable, or retractable. Fixed and removable bollards are passive barriers that are typically used along building perimeters or on sidewalks to prevent vehicles from them, while allowing pedestrians to pass them. In contrast to passive bollards, retractable bollards are active security barriers that can easily be raised and lowered to allow vehicles to pass between them. Thus, they can be used in driveways or on roads to control vehicular access. When the bollards are raised, they protect above the road surface and block the roadway; when they are lowered, they sit flush with the road surface, and thus allow traffic to pass over them. Retractable bollards are typically constructed from steel or other materials that have a low weight-to-volume ratio so that they require low power to raise and lower. Steel is also more resistant to breaking than is a more brittle material, such as concrete, and is better able to withstand direct vehicular impact without breaking apart.

Table 9.3. Pros and Cons of Gates

Pros	*Cons*
All three gate designs possess an intermediate crash rating, thereby allowing them to be utilized for medium to higher security applications. Requires very little maintenance. Can be tailored to blend in with perimeter fencing. Gate construction requires no roadway excavation. Cantilever gates are useful for roads with high crowns or drainage gutters. These types of barriers are extremely visible, which may deter intruders. Gates can also be used to control pedestrian traffic.	Gates have a slower response time (normally 10–15 seconds, but can be reduced to 7–10 seconds in emergency situations) than do other types of active security barriers, such as wedge barriers. Because emergency activation of the barrier causes more wear and tear on the system than does normal activation, it is recommended for use only in true emergency situations.

Source: USEPA, 2005.

Retractable bollards are installed in a trench dug across a roadway—typically at an entrance or gate. Installing retractable bollards requires preparing the road surface. Depending on the vendor, bollards can be installed either in a continuous slab of concrete, or in individual excavations with concrete poured in place. The required excavation for a bollard is typically slightly wider and slightly deeper than the bollard height when extended aboveground. The bottom of the excavation is typically lined with gravel to allow drainage. The bollards are then connected to a control panel which controls the raising and lowering of the bollards. Installation typically requires mechanical, electrical, and concrete work; if utility personnel with these skills are available, then the utility can install the bollards themselves. Table 9.4 lists the pros and cons of retractable bollards.

Table 9.4. Pros and Cons of Retractable Bollards

Pros	*Cons*
Bollards have a quick response time (normally 3 to 10 seconds, but can be reduced to 1–3 seconds in emergency situations). Bollards have an intermediate crash rating, which allows them to be utilized for medium to higher security applications.	Bollard installations will require construction below the surface of the roadway, which may interfere with buried utilities. Some maintenance is needed to ensure barrier is free to move up and down. The distance between bollards must be decreased (i.e., more bollards must be installed along the same perimeter) to make these systems effective against small vehicles (i.e., motorcycles).

Source: USEPA, 2005.

Portable/removable barriers, which can include removable crash beams and wedge barriers, are mobile obstacles that can be moved in and out of position on a roadway. For example, a crash beam may be completely removed and stored off-site when it is not needed. An additional example would be wedge barriers that are equipped with wheels that can be removed after the barricade is towed into place.

When portable barricades are needed, they can be moved into position rapidly. To provide them with added strength and stability, they are typically anchored to buttress boxes that are located on either side of the road. These buttress boxes, which may or may not be permanent, are usually filled with sand, water, cement, gravel, or concrete to make them heavy and aid in stabilizing the portable barrier. In addition, these buttresses can help dissipate any impact energy from vehicles crashing into the barrier itself.

Because these barriers are not anchored into the roadway, they do not require excavation or other related construction for installation. In contrast, they can be assembled and made operational in a short period of time. The primary shortcoming to this type of design is that these barriers may move if they are hit by vehicles. Therefore, it is important to carefully assess the placement and anchoring of these types of barriers to ensure that they can withstand the types of impacts that may be anticipated at that location. Table 9.5 lists the pros and cons of portable/removable barricades.

Because the primary threat to active security barriers is that vehicles will attempt to crash through them, their most important attributes are their size, strength, and crash resistance. Other important features for an active security barrier are the mechanisms by which the barrier is raised and lowered to allow authorized vehicle entry, and other factors, such as weather resistance and safety features.

Table 9.5. Pros and Cons of Portable/Removable Barricades

Pros	*Cons*
Installing portable barricades requires no foundation or roadway excavation. Can be moved in and out of position in a short period of time. Wedge barriers equipped with wheels can be easily towed into place. Minimal maintenance is needed to keep barriers fully operational.	Portable barriers may move slightly when hit by a vehicle, resulting in a lower crash resistance. Portable barricades typically require 7.75 to 16.25 seconds to move into place, and thus they are considered to have a medium response time when compared with other active barriers.

Source: USEPA, 2005.

PHYSICAL ASSET MONITORING AND CONTROL DEVICES—FLOATING BOAT BARRIER SYSTEM

A **Floating Boat Barrier System** consists of floating security devices that are installed upstream of a dam to prevent unauthorized waterborne entry. The brightly painted barriers provide a highly visible line of demarcation. These anti-intrusion barriers also protect against floating objects such as logs and debris.

ALARM SYSTEMS

An **alarm system** is a type of electronic monitoring system that is used to detect and respond to specific types of events—such as unauthorized access to an asset or a possible fire. These types of alarms are primarily integrated with process monitoring and reporting systems (i.e., SCADA systems). Note that this discussion does not focus on alarm systems that are not related to a facility's processes.

Alarm systems can be integrated with fire detection systems, intrusion detection systems (IDSs), access control systems, or CCTV systems, such that these systems automatically respond when the alarm is triggered. For example, a smoke detector alarm can be set up to automatically notify the fire department when smoke is detected or an intrusion alarm can automatically trigger cameras to turn on in a remote location so that personnel can monitor that location.

An alarm system consists of **sensors** that detect different types of events; an **arming station** that is used to turn the system on and off; a **control panel** that receives information, processes it, and transmits the alarm; and an **annunciator** that generates a visual and/or audible response to the alarm. When a sensor is tripped it sends a signal to a control panel, which triggers a visual or audible alarm and/or notifies a central monitoring station. A more complete description of each of the components of an alarm system is provided in the following.

Detection devices (also called **sensors**), are designed to detect a specific type of event (such as smoke, intrusion, etc.). Depending on the type of event they are designed to detect, sensors can be located inside or outside of the facility or other asset. When an event is detected, the sensors use some type of communication method (such as wireless radio transmitters, conductors, or cables) to send signals to the control panel to generate the alarm. For example, a smoke detector sends a signal to a control panel when it detects smoke.

Alarms use either **normally closed (NC)** or **normally open (NO)** electric loops, or "circuits," to generate alarm signals. In NC loops or circuits, all of the system's sensors and switches are connected in series. The contacts are "at rest" in the closed (on) position, and current continually passes through the system. However, when an event triggers the sensor, the loop is opened, breaking the flow of current through the system and triggering the alarm. NC switches are used more often than are NO switches because the alarm will be activated if the loop or circuit is broken or cut, thereby reducing the potential for circumventing the alarm. This is known as a "supervised" system.

In NO loops or circuits, all of the system's sensors and switches are connected in parallel. The contacts are "at rest" in the open (off) position, and no current passes through the system. However, when an event triggers the sensor, the loop is closed. This allows current to flow through the loop, powering the alarm. NO systems are not "supervised" because the alarm will not be activated if the loop or circuit is broken or cut. However, adding an end-of-line resistor to an NO loop will cause the system to alarm if tampering is detected.

An **arming station**, which is the main user interface with the security system, allows the user to arm (turn on), disarm (turn off), and communicate with the system. How a specific system is armed will depend on how it is used. For example, while IDSs can be armed for continuous operation (24 hours/day), they are usually armed and disarmed according to the work schedule at a specific location so that personnel going about their daily activities do not set off the alarms. In contrast, fire protection systems are typically armed 24 hours a day.

A **control panel** receives information from the sensors and sends it to an appropriate location, such as to a central operations station or to a 24-hour monitoring facility. Once the alarm signal is received at the central monitoring location, personnel monitoring for alarms can respond (such as by sending security teams to investigate or by dispatching the fire department).

An **annunciator** responds to the detection of an event by emitting a signal. This signal may be visual, audible, electronic, or a combination of these three. For example, fire alarm signals will always be connected to audible annunciators, whereas intrusion alarms may not be.

Alarms can be reported locally, remotely, or both locally and remotely. Local and remotely (centrally) reported alarms are discussed in more detail next.

A **local alarm** emits a signal at the location of the event (typically using a bell or siren). A "local only" alarm emits a signal at the location of the event but does not transmit the alarm signal to any other location (i.e., it does not transmit the alarm to a central monitoring location). Typically, the purpose of a "local only" alarm is to frighten away intruders, and possibly to attract the attention of someone who might notify the proper authorities. Because no signal is sent to a central monitoring location, personnel can only respond to a local alarm if they are in the area and can hear and/or see the alarm signal.

Fire alarm systems must have local alarms, including both audible and visual signals. Most fire alarm signal and response requirements are codified in the National Fire Alarm Code, National Fire Protection Association (NFPA) 72. NFPA 72 discusses the application, installation, performance, and maintenance of protective signaling systems and their components. In contrast to fire alarms, which require a local signal when fire is detected, many IDSs do not have a local alert device, because monitoring personnel do not wish to inform potential intruders that they have been detected. Instead, these types of systems silently alert monitoring personnel that an intrusion has been detected, thus allowing monitoring personnel to respond.

In contrast to systems that are set up to transmit "local only" alarms when the sensors are triggered, systems can also be set up to transmit signals to a **central location**, such as to a control room or guard post at the utility, or to a police or fire station. Most fire/smoke alarms are set up to signal both at the location of the event and at a fire station or central monitoring station. Many insurance companies require that facilities install certified systems that include alarm communication to a central station. For example, systems certified by the Underwriters Laboratory (UL) require that the alarm be reported to a central monitoring station.

The main differences between alarm systems lie in the types of event detection devices used in different systems. **Intrusion sensors**, for example, consist of two main categories: perimeter sensors and interior (space) sensors. **Perimeter intrusion sensors** are typically applied on fences, doors, walls, windows, and the like, and are designed to detect an intruder before he or she accesses a protected asset (i.e., perimeter intrusion sensors are used to detect intruders attempting to enter through a door, window). In contrast, **interior intrusion sensors** are designed to detect an intruder who has already accessed the protected asset (i.e., interior intrusion sensors are used to detect intruders once they are already within a protected room or building). These two types of detection devices can be complementary, and they are often used together to enhance security for an asset. For example, a typical intrusion alarm system might employ a perimeter glass-break detector that protects against intruders accessing a room through a window, as well as an ultrasonic interior sensor that detects intruders that have gotten into the room without using the window. Table 9.6 lists and describes types of perimeter and interior sensors.

Fire detection/fire alarm systems consist of different types of fire detection devices and fire alarm systems available. These systems may detect fire, heat, smoke, or a combination of any of these. For example, a typical fire alarm system might consist of heat sensors, which are located throughout a facility and which detect high temperatures or a certain change in temperature over a fixed time period. A different system might be outfitted with both smoke and heat detection devices. A summary of several different types of fire/smoke/heat detection sensors is provided in table 9.7.

Once a sensor in an alarm system detects an event, it must communicate an alarm signal. The two basic types of alarm communication systems are hardwired and wireless. Hardwired systems rely on wire that is run from the control panel to each of the detection devices and annunciators. Wireless systems transmit signals from a transmitter to a receiver through the air—primarily using radio or other waves. Hardwired systems are usually lower-cost, more reliable (they are not affected by terrain or environmental factors), and significantly easier to troubleshoot than are wireless systems. However, a major disadvantage of hardwired systems is that it may not be possible to hardwire all locations (e.g., it may be difficult to hardwire remote locations). In addition, running wires to their required locations can

Table 9.6. Perimeter and Interior Sensors

Type of Perimeter Sensor	*Description*
Foil	Foil is a thin, fragile, lead-based metallic tape that is applied to glass windows and doors. The tape is applied to the window or door, and electric wiring connects this tape to a control panel. The tape functions as a conductor and completes the electric circuit with the control panel. When an intruder breaks the door or window, the fragile foil breaks, opening the circuit and triggering an alarm condition.
Magnetic switches (reed switches)	The most widely used perimeter sensor. They are typically used to protect doors, as well as windows that can be opened (windows that cannot be opened are more typically protected by foil alarms).
Glass break detectors	Placed on glass and sense vibrations in the glass when it is disturbed. The two most common types of glass break detectors are shock sensors and audio discriminators.
Type of Interior Sensor	*Description*
Passive infrared (PIR)	Presently the most popular and cost effective interior sensors. PIR detectors monitor infrared radiation (energy in the form of heat) and detect rapid changes in temperature within a protected area. Because infrared radiation is emitted by all living things, these types of sensors can be very effective.
Quad PIRs	Consist of two dual-element sensors combined in one housing. Each sensor has a separate lens and a separate processing circuitry, which allows each lens to be set up to generate a different protection pattern
Ultrasonic detectors	Emit high frequency sound waves, and sense movement in a protected area by sensing changes in these waves. The sensor emits sound waves that stabilize and set a baseline condition in the area to be protected. Any subsequent movement within the protected area by a would-be intruder will cause a change in these waves, thus creating an alarm condition.
Microwave detectors	Emit ultra-high frequency radio waves, and the detector senses any changes in these waves as they are reflected throughout the protected space. Microwaves can penetrate through walls, and thus a unit placed in one location may be able to protect multiple rooms.
Dual technology devices	Incorporate two different types of sensor technology (such as PIR and microwave technology) together in one housing. When both technologies sense an intrusion, an alarm is triggered.

Source: USEPA, 2005.

Table 9.7. Fire/Smoke/Heat Detection Sensors

Detector Type	*Description*
Thermal detector	Sense when temperatures exceed a set threshold (fixed temperature detectors) or when the rate of change of temperature increases over a fixed time period (rate-of-rise detectors).
Duct detector	Is located within the heating and ventilation ducts of the facility. This sensor detects the presence of smoke within the system's return or supply ducts. A sampling tube can be added to the detector to help span the width of the duct.
Smoke detectors	Sense invisible and/or visible products of combustion. The two principle types of smoke detectors are photoelectric and ionization detectors. The major differences between these devices are: • Photoelectric smoke detectors react to visible particles of smoke. These detectors are more sensitive to the cooler smoke with large smoke particles that is typical of smoldering fires. • Ionization smoke detectors are sensitive to the presence of ions produced by the chemical reactions that take place with few smoke particle, such as those typically produced by fast burning/flaming fires.
Multisensor detectors	Are a combination of photoelectric and thermal detectors. The photoelectric sensor serves to detect smoldering fires, while the thermal detector senses the eat give off from fast burning/flaming fires.
Carbon monoxide (CO) detectors	Are used to indicate the outbreak of fire by sensing the level of carbon monoxide in the air. The detector has an electrochemical cell which senses carbon monoxide, but not some or other products of combustion.
Beam detectors	Are designed to protect large, open spaces such as industrial warehouses. These detectors consist of three parts: the transmitter, which projects a beam of infrared light; the receiver, which registers the light and produces an electrical signal; and the interface, which processes the signal and generates alarm of fault signals. In the event of a fire, smoke particles obstruct the beam of light. Once a preset threshold is exceeded, the detector will go into alarm.
Flame detectors	Sense either ultraviolet (UV) or infrared (IR) radiation emitted by a fire.
Air-sampling detectors	Actively and continuously sample the air from a protected space and are able to sense the pre-combustion stages of incipient fire.

Source: USEPA, 2005.

be both time consuming and costly. The major advantage to using wireless systems is that they can often be installed in areas where hardwired systems are not feasible. However, wireless components can be much more expensive when compared to hardwired systems. In addition, in the past, it has been difficult to perform self-diagnostics on wireless systems to confirm that they are communicating properly with the controller. Presently, the majority of wireless systems incorporate supervising circuitry, which allows the subscriber to know immediately if there is a problem with the system (such as a broken detection device or a low battery), or if a protected door or window has been left open.

BIOMETRIC SECURITY SYSTEMS

Biometrics involves measuring the unique physical characteristics or traits of the human body. Any aspect of the body that is measurably different from person to person—for example fingerprints or eye characteristics—can serve as a unique biometric identifier for that individual. Biometric systems recognizing fingerprints, palm shape, eyes, face, voice, and signature comprise the bulk of the current biometric systems that recognize other biological features do exist.

Biometric security systems use biometric technology combined with some type of locking mechanisms to control access to specific assets. In order to access an asset controlled by a biometric security system, an individual's biometric trait must be matched with an existing profile stored in a database. If there is a match between the two, the locking mechanisms (which could be a physical lock, such as at a doorway, an electronic lock, such as at a computer terminal, or some other type of lock) is disengaged, and the individual is given access to the asset.

A biometric security system is typically comprised of the following components:

- A sensor, which measures/records a biometric characteristic or trait
- A control panel, which serves as the connection point between various system components. The control panel communicates information back and forth between the sensor and the host computer, and controls access to the asset by engaging or disengaging the system lock based on internal logic and information from the host computer

- A host computer, which processes and stores the biometric trait in a database
- Specialized software, which compares an individual image taken by the sensor with a stored profile or profiles
- A locking mechanism which is controlled by the biometric system
- A power source to power the system.

Biometric Hand and Finger Geometry Recognition

Hand and finger geometry recognition is the process of identifying an individual through the unique "geometry" (shape, thickness, length, width, etc.) of that individual's hand or fingers. Hand geometry recognition has been employed since the early 1980s and is among the most widely used biometric technologies for controlling access to important assets. It is easy to install and use, and is appropriate for use in any location requiring use of two finger highly accurate, nonintrusion biometric security. For example, it is currently used in numerous workplaces, daycare facilities, hospitals, universities, airports, refineries, and power plants.

A newer option within hand geometry recognition technology is finger geometry recognition (not to be confused with fingerprint recognition). Finger geometry recognition relies on the same scanning methods and technologies as does hand geometry recognition, but the scanner only scans two of the user's fingers, as opposed to his entire hand. Finger geometry recognition has been in commercial use since the mid-1990s and is mainly used in time and attendance applications (i.e., to track when individuals have entered and exited a location). To date the only large-scale commercial use of two-finger geometry for controlling access is at Disney World, where season pass holders use the geometry of their index and middle finger to gain access to the facilities.

To use a hand or finger geometry unit, an individual presents his or her hand or fingers to the biometric unit for "scanning." The scanner consists of a Charged Coupled Device (CCD), which is essentially a high-resolution digital camera; a reflective platen on which the hand is placed; and a mirror or mirrors that help capture different angles of the hand or fingers. The camera "scans" individual geometric characteristics of the hand or fingers by taking multiple images while the user's hand rests on the reflective platen. The camera also captures "depth," or three-dimensional information, through light reflected from the mirrors and the reflective platen. This live image is then compared to a "template" that was previously established for that individual when they were enrolled in the system. If the live scan of the individual matches the stored template, the individual is "verified," and is given access to that asset. Typically, verification takes about two seconds. In access control applications, the scanner is usually connected to some sort of electronic lock, which unlocks the door, turnstile, or other entry barrier when the user is verified. The user can then proceed through the entrance. In time and attendance applications, the time that an individual checks in an out of a location is stored for later use.

As discussed, hand and finger geometry recognition systems can be used in several different types of applications, including access control and time and attendance tracking. While time and attendance tracking can be used for security, it is primarily used for operations and payroll purposes (i.e., clocking in and clocking out). In contrast, access control applications are more likely to be security related. Biometric systems are widely used for access control, and can be used on various types of assets, including entryways, computers, vehicles, etc. However, because of their size, hand/finger recognition systems are primarily used in entryway access control applications.

Biometric Iris Recognition

The iris, which is the colored or pigmented area of the eye surrounded by the sclera (the white portion of the eye), is a muscular membrane that controls the amount of light entering the eye by contracting or expanding the pupil (the dark center of the eye). The dense, unique patterns of connective tissue in the human iris were first noted in 1936, but it was not until 1994, when algorithms for iris recognition were created and patented, that commercial applications using biometric iris recognition began to be used extensively. There are now two vendors producing iris recognition technology: both the original developer of these algorithms, as well as a second company, which has developed and patented a different set of algorithms for iris recognition.

The iris is an ideal characteristic for identifying individuals because it is formed *in utero*, and its unique patterns stabilize around eight months after birth. No two irises are alike; neither an individual's right or left irises, nor the irises of identical twins. The iris is protected by the cornea (the clear covering over the eye), and therefore it is not subject to the aging or physical changes (and potential variation) that are common to some other biometric measures, such as the hand, fingerprints, and the face. Although some limited changes can occur naturally over time, these changes generally occur in the iris' melanin and therefore affect only the eye's color, and not its unique patterns (in addition, because iris scanning uses only black-and-white images, color changes would not affect the scan anyway). Thus, barring specific injuries or certain rate surgeries directly affecting the iris, the iris' unique patterns remain relatively unchanged over an individual's lifetime.

Iris recognition systems employ a monochromatic or black-and-white video camera that uses both visible and near infrared light to take video of an individual's iris. Video is used rather than still photography as an extra security procedure. The video is used to confirm the normal continuous fluctuations of the pupil as the eye focuses, which ensures that the scan is of a living human being, and not a photograph or some other attempted hoax. A high resolution image of the iris is then captured or extracted from the video, using a device often referred to as a "frame grabber." The unique characteristics identified in this image are then converted into a numeric code, which is stored as a "template" for that user.

CARD IDENTIFICATION/ACCESS/TRACKING SYSTEMS

A card reader system is a type of electronic identification system that is used to identify a card and then perform an action associated with that card. Depending on the system, the card may identify where a person is or where they were at a certain time; or it may authorize another action, such as disengaging a lock. For example, a security guard may use his card at card readers located throughout a facility to indicate that he has checked a certain location at a certain time. The reader will store the information and/or send it to a central location, where it can be checked later to ensure that the guard has patrolled the area. Other card reader systems can be associated with a lock, so that the cardholder must have their card read and accepted by the reader before the lock disengages.

A complete card reader system typically consists of the following components:

- Access cards that are carried by the user
- Card readers, which read the card signals and send the information to control units
- Control units, which control the response of the card reader to the card
- A power source

A "card" may be a typical card or another type of device, such as a key fob or wand. These cards store electronic information, which can range from a simple code (i.e., the alphanumeric code on a Proximity card) to individualized personal data (i.e., biometric data on a Smartcard). The card reader reads the information stored on the card and sends it to the control unit, which determines the appropriate action to take when a card is presented. For example in a card access system, the control unit compares the information on the card vs. stored access authorization information to determine if the card holder is authorized to proceed through the door. If the information stored in the card reader system indicates that the key is authorized to allow entrance through the doorway, the system disengages the lock and the keyholder can proceed through the door.

There are many different types of card reader systems on the market. The primary differences between card reader systems are different in the way that data is encoded on the cards and in the way these data are transferred between the card and the card reader, and in the types of applications for which they are best suited. However, all card systems are similar in the way that the card reader and control unit interact to respond to the card.

While card readers are similar in the way that the card reader and control unit interact to control access, they are different in the way data is encoded on the cards and the way these data are transferred between the card and the card reader. There are several types of technologies available for card reader systems. These include:

- Proximity
- Wiegand
- Smartcard
- Magnetic Stripe
- Barcode

- Infrared
- Barium Ferrite
- Hollerith
- Mixed Technologies

Table 9.8 below summarizes various aspects of card reader technologies. The determination for the level of security rate (low, moderate, or high) based on the level of technology a given card reader system has and how simple it is to duplicate that technology, and thus bypass the security. Vulnerability ratings were based on whether the card reader can be damaged easily due to frequent use or difficult working conditions (i.e., weather conditions if the reader is located outside). Often this is influenced by the number of moving parts in the system—the more moving parts, then greater the system's potential susceptibility to damage. The life cycle rating is based on the durability of a given card reader system over its entire operational period. Systems requiring frequent physical contact between the reader and the card often have a shorter life cycle due to the wear and tear to which the equipment is exposed. For many card reader systems, the vulnerability rating and life cycle ratings have a reciprocal relationship. For instance, if a given system has a high vulnerability rating it will almost always have a shorter life cycle.

Card reader technology can be implemented for facilities of any size and with any number of users. However, because individual systems vary in the complexity of their technology and in the level of security they can provide to a facility, individual users must determine the appropriate system for their needs. Some important features to consider when selecting a card reader system include:

- The technological sophistication and security level of the card system
- The size and security needs of the facility
- The frequency with which the card system will be used. For systems that will experience a high frequency of use it is important to consider a system that has a longer life cycle and lower vulnerability rating, thus making it more cost effective to implement.
- The conditions in which the system will be used (i.e., will it be used on the interior or exterior of buildings, does it require light or humidity controls).

Table 9.8. Card Reader Technology

Types of Card Readers	*Technology*	*Life Cycle*	*Vulnerability*	*Level of Security*
Proximity	Embedded radio frequency circuits encoded with unique information	Long	Virtually none	Moderate–High
Wiegand	Short lengths of small-diameter, special alloy wire with unique magnetic properties	Long	Low susceptibility to damage; high durability due to embedded wires	Moderate–Expensive
Magnetic Stripe	Electromagnetic charges to encode information on tape attached to back of card	Moderate	Moderately susceptible to damage due to frequency of use	Low–Moderate
Bar Code	Series of narrow and wide bars and spaces	Short	High; easily damaged	Low
Hollerith	Holes punched in a plastic or paper card and read optically	Short	High; easily damaged from frequent use	Low
Infrared	An encoded shadow pattern within the card, read using an infrared scanner	Moderate	IR scanners are optical and thus, vulnerable to contamination	High
Barium Ferrite	Uses small bits of magnetized barium ferrite, placed inside a plastic card. The polarity and location of the "spots" determines the coding	Moderate	Low susceptibility to damage; durable since spots are embedded in the material	Moderate–High
Smartcards	Patterns or series of narrow and wide bars and spaces	Short	High susceptibility to damage, low durability	Highest

Source: USEPA, 2005.

Most card reader systems can operate under normal environmental conditions, and therefore this would be a mitigating factor only in extreme conditions

- System costs

EXTERIOR INTRUSION SENSORS

An **exterior intrusion sensor** is a detection device that is used in an outdoor environment to detect intrusions into a protected area. These devices are designed to detect an intruder, and then communicate an alarm signal to an alarm system. The alarm system can respond to the intrusion in many different ways, such as by triggering an audible or visual alarm signal, or by sending an electronic signal to a central monitoring location that notifies security personnel of the intrusion.

Intrusion sensor can be used to protect many kinds of assets. Intrusion sensors that protect physical space are classified according to whether they protect indoor, or "interior" space (i.e., an entire building or room within a building), or outdoor, or "exterior" space (i.e., a fence line or perimeter). Interior intrusion sensors are designed to protect the interior space of a facility by detecting an intruder who is attempting to enter, or who has already entered a room or building. In contrast, exterior intrusion sensors are designed to detect an intrusion into a protected outdoor/exterior area. Exterior protected areas are typically arranged as zones or exclusion areas placed so that the intruder is detected early in the intrusion attempt before the intruder can gain access to more valuable assets (e.g., into a building located within the protected area). Early detection creates additional time for security forces to respond to the alarm.

Exterior intrusion sensors are classified according to how the sensor detects the intrusion within the protected area. The three classes of exterior sensor technology include:

- Buried line sensors
- Fence-associated sensors
- Freestanding sensors

Buried-Line Sensors

As the name suggests, buried line sensors are sensors that are buried underground, and are designed to detect disturbances within the ground—such as disturbances caused by an intruder digging, crawling, walking, or running on the monitored ground. Because they sense ground disturbances, these types of sensors are able to detect sense able to detect intruder activity both on the surface and below ground. Individual types of exterior buried line sensors function in different ways, including, by detecting motion, pressure, or vibrations within the protected ground, or by detecting changes in some type of field (e.g., magnetic field) that the sensors generate within the protected ground. Specific types of buried line sensors include **pressure or seismic** sensors, **magnetic field** sensors, **ported coaxial cables**, and **fiber-optic cables.** The four types of sensors are described in more detail below. Table 9.9 presents the distinctions between the four types of buried sensors.

Table 9.9. Types of Buried Sensors

Type	*Description*
Pressure or Seismic	Responds to disturbances in the soil.
Magnetic Field	Responds to a change in the local magnetic field caused by the movement of nearby metallic material.
Ported Coaxial Cables	Responds to motion of a material with a high dielectric constant or high conductivity near the cables.
Fiber-Optic Cables	Responds to a change in the shape of the fiber that can be sensed using sophisticated sensors and computer signal processing.

Source: Adapted from Garcia, M.L., 2001.

- **Buried-line pressure or seismic sensors** detect physical disturbances to the ground—such as vibrations or soil compression—caused by intruders walking, driving, digging, or otherwise physically contacting the protected ground. These sensors detect disturbances from all directions and, therefore, can protect an area radially outward from their location; however, because detection may weaken as a function of distance from the disturbance, choosing the correct burial depth from the design area will be crucial. In general, sensors buried at a shallow depth protect a relatively small area but have a high probability of detecting intrusion within that area, while sensors buried at a deeper depth protect a wider area but have a lower probability of detecting intrusion into that area.
- **Buried line magnetic field sensors** detect changes in a local magnetic field that are caused by the moment of metallic objects within that field. This type of sensor can detect ferric metal objects worn or carried by an intruder entering a protected area on foot as well as vehicles being driven into the protected area.

- **Buried line ported coaxial cable sensors** detect the motion of any object (i.e., human body, metal) possessing high conductivity and located within close proximity to the cables. An intruder entering into the protected space creates an active disturbance in the electric field, thereby triggering an alarm condition.
- **Buried line fiber-optic cable sensors** detect changes in the attenuation of light signals transmitted within the cable. When the soil around the cable is compressed, the cable is distorted, and the light signal transmitted through the cable changes, initiating an alarm. This type of sensor is easy to install because it can be buried at a shallow burial depth (only a few centimeters) and still be effective.

Fence-Associated Sensors

Fence-associated sensors are either attached to an existing fence, or are installed in such a way as to create a fence. These sensors detect disturbances to the fence—such as those caused by an intruder attempting to climb the fence or by an intruder attempting to cut or lift the fence fabric. Exterior fence-associated sensors include fence-disturbance sensors, taut-wire sensor fences, and electric field or capacitance sensors. Details on each of these sensor types are provided next.

- **Fence-disturbance sensors** detect the motion or vibration of a fence, such as that caused by an intruder attempting to climb or cut through the fence. In general, fence disturbance sensors are used on chain link fences or on other fence types where a moveable fence fabric is hung between fence posts.
- **Taut-wire sensor fences** are similar to fence-disturbance sensors except that instead of attaching the sensors to a loose fence fabric, the sensors are attached to a wire that is stretched tightly across the fence. These types of systems are designed to detect changes in the tension of the wire rather than vibrations in the fence fabric. Taut-wire sensor fences can be installed over existing fences, or as stand-alone fence systems.
- **Electric field or capacitance sensors** detect changes in capacitive coupling between wires that are attached to, but electrically isolated from, the fence. As opposed to other fence-associated intrusion sensors, both electric field and capacitance sensors generate an electric field that radiates out from

the fence line, resulting in an expanded zone of protection relative to other fence-associated sensors, and allowing the sensor to detect an intruders' presence before they arrive at the fence line. Note: proper spacing is necessary during installation of the electric field sensor to detect a would-be intruder from slipping between largely spaced wires.

Free-Standing Sensors

These sensors, which include active infrared, passive infrared, bistatic microwave, monostatic microwave, dual-technology, and video motion detection (VMD) sensors, consist of individual sensor units or components that can be set up in a variety of configurations to meet a user's needs. They are installed above-ground and, depending on how they are oriented relative to each other, can be used to establish a protected perimeter or a protected space. More details on each of these sensor types are provided next.

- **Active infrared sensors** transmit infrared energy into the protected space, and monitor for changes in this energy caused by intruders entering that space. In a typical application, an infrared light beam is transmitted from a transmitter unit to a receiver unit. If an intruder crosses the beam, the beam is blocked, and the receiver unit detects a change in the amount of light received, triggering an alarm. Different sensors can see single- and multiple-beam arrays. Single-beam infrared sensors transmit a single infrared beam. In contrast, multiple-beam infrared sensors transmit two or more beams parallel to each other. This multiple-beam sensor arrangement creates an infrared "fence."
- **Passive infrared (PIR) sensors** monitor the ambient infrared energy in a protected area, and evaluate changes in that ambient energy that may be caused by intruders moving through the protected area. Detection ranges can exceed 100 yards on cold days with size and distance limitations dependent upon the background temperature. PIR sensors generate a non-uniform detection pattern (or "curtain") that has areas (or "zones") of more sensitivity and areas of less sensitivity. The specific shape of the protected area is determined by the detector's lenses. The general shape common to many detection patterns is a series of long "fingers" emanating from the PIR and spreading in various directions. When intruders enter the detec-

tion area, the PIR sensor detects differences in temperature due to the intruder's body heat and triggers an alarm. While the PIR leaves unprotected areas between its fingers, an intruder would be detected if he or she passed from a nonprotected area to a protected area.

- **Microwave sensors** detect changes in received energy generated by the motion of an intruder entering into a protected area. Monostatic microwave sensors incorporate transmitter and a receiver in one unit, while bistatic sensors separate the transmitter and the receiver into different units. Monostatic sensors are limited to a coverage area of 400 feet, while bistatic sensors can cover an area up to 1,500 feet. For bistatic sensors, a zone of no detection exists in the first few feet in front of the antennas. This distance from the antennas to the point at which the intruder is first detected is known as the offset distance. Due to this offset distance, antennas must be configured so that they overlap one another (as opposed to being adjacent to each other), thereby creating long perimeters with a continuous line of detection.
- **Dual-technology sensors** consist of two different sensor technologies incorporated together into one sensor unit. For example, a dual technology sensor could consist of a passive infrared detector and a monostatic microwave sensor integrated into the same sensor unit.
- **VMD sensors** monitor video images from a protected area for changes in the images. Video cameras are used to detect unauthorized intrusion into the protected area by comparing the most recent image against a previously established one. Cameras can be installed on towers or other tall structures so that they can monitor a large area.

FENCES

A fence is a physical barrier that can be set up around the perimeter of an asset. Fences often consist of individual pieces (such as individual pickets in a wooden fence, or individual sections of a wrought iron fence) that are fastened together. Individual sections of the fence are fastened together using posts, which are sunk into the ground to provide stability and strength for the sections of the fence hung between them. Gates are installed between individual sections of the fence to allow access inside the fenced area.

Many fences are used as decorative architectural features to separate physical spaces for each other. They may also be used to physically mark the location of a boundary (such as a fence installed along a properly line). However, a fence can also serve as an effective means for physically delaying intruders from gaining access to a Dam Sector asset. For example, many utilities install fences around their primary facilities, around remote pump stations, or around hazardous petrochemical materials storage areas or sensitive areas within a facility. Access to the area can be controlled through security at gates or doors through the fence (e.g., by posting a guard at the gate or by locking it). In order to gain access to the asset, unauthorized persons could either have to go around or through the fence.

Fences are often compared with walls when determining the appropriate system for perimeter security. While both fences and walls can provide adequate perimeter security, fences are often easier and less expensive to install than walls. However, they do not usually provide the same physical strength that walls do. In addition, many types of fences have gaps between the individual pieces that make up the fence (i.e., the spaces between chain links in a chain link fence). Thus, many types of fences allow the interior of the fenced area to be seen. This may allow intruders to gather important information about the locations or defenses of vulnerable areas within the facility.

Important security attributes of a fence include the height to which it can be constructed, the strength of the material comprising the fence, the method and strength of attaching the individual sections of the fence together at the posts and the fence's ability to restrict the view of the assets inside the fence. Additional considerations should include the ease of installing the fence and the ease of removing and reusing sections of the fence. Table 9.10 provides a comparison of the important security and usability features of various fence types.

Table 9.10. Comparison of Different Fence Types

Specifications	*Chain Link*	*Iron*	*Wire (Wirewall)*	*Wood*
Height limitations	12'	12'	12'	8'
Strength	Medium	High	High	Low
Installation Requirements	Low	High	High	Low
Ability to Remove/Reuse	Low	High	Low	High
Ability to Replace/Repair	Medium	High	Low	High

Source: USEPA, 2005.

Some fences can include additional measures to delay, or even detect, potential intruders. Such measures may include the addition of barbed wire, razor wire, or other deterrents at the top of the fence. Barbed wire is sometimes employed at the base of fences as well. This can impede a would-be intruder's progress in even reaching the fence. Fences may also be fitted with security cameras to provide visual surveillance of the perimeter. Finally, some facilities have installed motion sensors along their fences to detect movement on the fence. Several manufacturers have combined these multiple perimeter security features into one product and offer alarms, and other security features.

The correct implementation of a fence can make it a much more effective security measure. Security experts recommend the following when a facility constructs a fence:

- The fence should be at least 7–9 feet high.
- Any outriggers, such as barbed wire, that are affixed on top of the fence should be angled out and away from the facility, and not in towards the facility. This will make climbing the fence more difficult, and will prevent ladders from being placed against the fence.
- Other types of hardware can increase the security of the fence. This can include installing concertina wire along the fence (this can be done in front of the fence or at the top of the fence), or adding intrusion sensors, camera, or other hardware to the fence.
- All undergrowth should be cleared for several feet (typically 6 feet) on both sides of the fence. This will allow for a clearer view of the fence by any patrols in the area.
- Any trees with limbs or branches hanging over the fence should be trimmed so that intruders cannot use them to go over the fence. Also, it should be noted that fallen trees can damage fences, and so management of trees around the fence can be important. This can be especially important in areas where fence goes through a remote area.
- Fences that do not block the view from outside the fence to inside the fence allow patrols to see inside the fence without having to enter the facility.
- "No Trespassing" signs posted along the fence can be a valuable tool in prosecuting any intruders who claim that the fence was broken and that they did not enter through the fence illegally. Adding signs that highlight the local ordinances against trespassing can further persuade simple troublemakers for illegally jumping/climbing the fence. Electrical substation and other electrical component installations should have clearly visible signage warning of HIGH VOLTAGE and the dangers of electrical shock.

FILMS FOR GLASS SHATTER PROTECTION

Most Dam Sector utilities have numerous windows on the outside of buildings, in doors, and in interior offices. In addition, many facilities have glass doors or other glass structures, such as glass walls or display cases. These glass objects are potentially vulnerable to shattering when heavy objects are thrown or launched at them, when explosions occur near them, or when there are high winds (for exterior glass). If the glass is shattered, intruders may potentially enter an area. In addition, shattered glass projected into a room from an explosion or from an object being thrown through a door or window can injure and potentially incapacitate personnel in the room. Materials that prevent glass from shattering can help to maintain the integrity of the door, window, or other glass object, and can delay an intruder from gaining access. These materials can also prevent flying glass and thus reduce potential injuries.

Materials designed to prevent glass from shattering include specialized films and coatings. These materials can be applied to existing glass objects to improve their strength and their ability to resist shattering. The films have been tested against many scenarios that could result in glass breakage, including penetration by blunt objects, bullets, high winds, and simulated explosions. Thus, the films are tested against both simulated weather scenarios (which could include both the high winds themselves and the force of objects blown into the glass), as well as more criminal/terrorist scenarios where the glass, or is subject to explosives or bullets. Many vendors provide information on the results of these types of tests, and thus potential users can compare different product lines to determine which products best suit their needs.

The primary attributes of films for shatter protection are:

- The materials from which the film is made
- The adhesive that bonds the film to the glass surface
- The thickness of the film

Standard glass safety films are designed from high strength polyester. Polyester provides both strength and elasticity, which is important in absorbing the impact of an object, spreading the force of the impact over the entire film, and resisting tearing. The polyester is also designed to be resistant to scratching, which can result when films are cleaned with abrasives or other industrial cleaners.

The bonding adhesive is important in ensuring that the film does not tear away from the glass surface. This can be especially important when the glass is broken, so that the film does not peel off the glass and allow it to shatter. In addition, films applied to exterior windows can be subject to high concentrations of UV light, which can break down bonding materials.

Film thickness is measured in gauge or mils. According to test results reported by several manufacturers, film thickness appears to affect resistance to penetration/tearing, with thicker films being more resistant to penetration and tearing. However, the appreciation of a thicker film did not decrease glass fragmentation.

Many manufacturers offer films in different thicknesses. The "standard" film is usually one 4 mil layer; thicker films are typically composed of several layers of the standard 4 mil sheet (1 mil equals 0.001 inch). However, newer technologies have allowed the polyester to be "microlayered" to produce a stronger film without significantly increasing its thickness. In this microlayering process, each laminate film is composed of multiple micro-thin layers of polyester woven together an alternating angles. This provides increased strength for the film, while maintaining the flexibility and thin profile of one film layer.

As described, many vendors test their products in various scenarios that would lead to glass shattering, including simulated bomb blasts and simulation of the glass being struck by wind-blown debris. Some manufacturers refer to the Government Services Administration standard for bomb blasts, which require resistance to tearing for a 4 PSI blast. Other manufacturers use other measures and test for resistance to tearing. Many of these tests are not "standard," in that no standard testing or reporting methods have been adopted by any of the accepted standards-setting institutions. However, many of the vendors publish the procedure and the results of these tests on their websites, and this may allow users to evaluate the protectiveness of these films. For example, several vendors evaluate the "protectiveness" of their films and the "hazard" resulting from blasts near windows with and without protective films. Protectiveness is usually evaluated based on the percentage of glass ejected from the window and the height at which that ejected glass travels during the blast (e.g., if the blasted glass tends to project upward in to a room—potentially toward people's faces—it is a higher hazard than if it is blown downward into the room toward people's feet). There are some standard measures of glass breakage. For example, several vendors indicated that their products exceed the American Society for Testing and Materials (ASTM) standard 64Z-95 "Standard Test Method for Glazing and Glazing Systems Subject to Air Blast Loadings." Vendors often compare the results of some sort of penetration or force test, ballistic tests, or simulated explosions with unprotected glass versus glass onto which their films have been applied. Results generally show that applying films to the glass surfaces reduces breakage/penetration of the glass and can reduce the amount and direction of glass ejected from the frame. This in turn reduces the hazard from flying glass.

In addition to these types of tests, many vendors conduct standard physical tests on their products, such as tests for tensile strength and peel strength. Tensile strength indicates the strength per area of material, while the peel strength indicates the force it would take to peel the product from the glass surface. Several vendors indicate that their products exceed American National Standards Institute (ANSI) standard Z97.1 for tensile strength and adhesion.

Vendors typically have a warranty against peeling or other forms of deterioration of their products. However, the warranty requires that the films be installed by manufacturer-certified technicians to ensure that they are applied correctly and therefore that the warranty is in effect. Warranties from different manufacturers may vary. Some may cover the cost of replacing the material only, while others include material plus installation. Because installation costs are significantly greater than material costs, different warranties may represent a large difference in potential costs.

FIRE HYDRANT LOCKS

Fire hydrants are installed at strategic locations throughout a community's water distribution system to supply water for firefighting. However, because there are many hydrants in a system and they are often located in residential neighborhood, industrial districts, and other areas where they cannot be easily observed and/or guarded, they are potentially vulnerable to unauthorized access. Many municipalities, states, and EPA Regions have recognized this potential vulnerability and have instituted programs to lock hydrants. For example, EPA Region 1 has included locking hydrants as number 7 on its "Drinking Water Security and Emergency Preparedness" Top Ten List for small groundwater suppliers.

A "hydrant lock" is a physical security device designed to prevent unauthorized access to the water supply through a hydrant. They can also ensue water and water pressure availability to fire fighters and prevent water theft and associated lost water revenue. These locks have been successfully used in numerous municipalities and in various climates and weather conditions.

Fire hydrant locks are basically steel covers or caps that are locked in place over the operating nut of a fire hydrant. The lock prevents unauthorized persons from accessing the operating nut and opening the fire hydrant valve. The lock also makes it more difficult to remove the bolts from the hydrant and access the system that way. Finally, hydrant locks shield the valve from being broken off. Should a vandal attempt to breach the hydrant lock by force and succeed in breaking the hydrant lock, the vandal will only succeed in bending the operating valve. If the hydrant's operating valve is bent, the hydrant will not be operational, but the water asset remains protected and inaccessible to vandals. However, the entire hydrant will need to be replaced.

Hydrant locks are designed so that the hydrants can be operated by special "key wrenches" without removing the lock. These specialized wrenches are generally distributed to the fire department, the public works department, and other authorized persons so that they can access the hydrants as needed. An inventory of wrenches and their serial numbers is generally kept by a municipality so that the location of all wrenches is known. These operating key wrenches may only be purchased by registered lock owners.

The most important features of hydrant are their strength and the security of their locking systems. The locks must be strong so that they cannot be broken off. Hydrant locks are constructed from stainless or alloyed steel. Stainless steel locks are stronger and are ideal for all climates; however, they are more expensive than alloy locks. The locking mechanisms for each fire hydrant locking system ensure that the hydrant can only be operated by authorized personnel who have the specialized key to work the hydrant.

HATCH SECURITY

A hatch is basically a door that is installed on a horizontal plane (such as in a floor, a paved lot, or a ceiling), instead of on a vertical plane (such as in a building wall). Hatches are usually used to provide access to assets that are either located underground (such as hatches to basements or underground vaults and storage areas), or to assets located above ceilings (such as emergency roof exits). At chemical industrial facilities, hatches are typically used to provide access to underground vaults containing pumps, meter chambers, valves, or piping, or to the interior of chemical tanks or covered water reservoirs. Securing a hatch by locking it or upgrading materials to give the hatch added strength can help to delay unauthorized access to any asset behind the hatch.

Like all doors, a hatch consists of a frame anchored to the horizontal structure, a door or doors, hinges connecting the door/doors to the frame, and a latching or locking mechanism that keeps the hatch door/doors closed.

It should be noted that improving hatch security is straightforward, and that hatches with upgraded security features can be installed new or they can be retrofit for existing applications.

Depending on the application, the primary security-related attributes of a hatch are the strength of the door and frame, its resistance to the elements and corrosion, it ability to be sealed against water or gas, and its locking features.

Hatches must be both strong and lightweight so that they can withstand typical static loads (such as people or vehicles walking or driving over them) while still being easy to open.

In addition, because hatches are typically installed at outdoor locations, they are usually designed from corrosion-resistant metal that can withstand the elements. Therefore, hatches are typically constructed from high gauge steel or lightweight aluminum.

Aluminum is typically the material of choice for hatches because it is lightweight and more corrosion resistant relative to steel. Aluminum is not as rigid as steel, so aluminum hatch doors may be reinforced with aluminum stiffeners to provide extra strength and rigidity. The doors are usually constructed from single or double layers (or "leaves") of material. Single-leaf designs are standard for smaller hatches, while double leaf deigns are required for larger hatches. In addition, aluminum products do not require painting. This reflected in the warranties available with different products. Product warranties range from 10 years to lifetime.

Steel is heavier per square foot than aluminum, and thus steel hatches will be heavier and more difficult to open than aluminum hatches of the same size. However, heavy steel hatch doors may have spring-loaded, hydraulic, or gas openers or other specialized features that help in opening the hatch and in keeping it open.

Many hatches are installed in outdoor areas, often in roadways or pedestrian areas. Therefore, the hatch installed for any given application must be designed to withstand the expected load at that location. Hatches are typically solid to withstand either pedestrian of vehicle loading. Pedestrian loading hatches are typically designed to withstand either 150 or 300 pounds per square feet (psf) of loading. The vehicle loading standard is the American Association of State Highway and Transportation Officials (AASHTO) H-20 wheel loading standard of 16,000 pounds over an 8-inch by 20-inch area. It should be noted that these design parameters are for static loads and not dynamic loads; thus, the loading capabilities may not reflect potential resistance to other types of loads that may be more typical of an intentional threat, such as repeated blows from a sledge hammer or pressure generated by bomb blasts or bullets.

The typical design for a watertight hatch includes a channel frame that directs water away from the hatch. This can be especially important in a hatch on a storage tank because this will prevent liquid contaminants from being dumped on the hatch and leaking through into the interior. Hatches can also be constructed with gasket seals that are air-, odor-, and gas-tight.

Typically, hatches for pedestrian loading applications have hinges located on the exterior of the hatch, while hatches designed for H-20 loads have hinges located in the interior of the hatch. Hinges located on the exterior of the hatch may be able to be removed; thereby allowing intruders to remove the hatch door and access the asset behind the hatch. Therefore, installing H-20 hatches even for applications which do not require H-20 loading levels may increase security, because intruders will not be able to tamper with the hinges and circumvent the hatch this way.

In addition to the location of the hinges, stock hinges can be replaced with heavy duty or security hinges that are more resistant to tampering.

The hatch locking mechanism is perhaps the most important part of hatch security. There are a number of locks that can be implemented for hatches, including:

- Slam locks (internal locks that are located within the hatch frame)
- Recessed cylinder locks
- Bolt locks
- Padlocks

LADDER ACCESS CONTROL

Dam Sector facilities have a number of assets that are raised above ground level, including electrical substations, pumping stations, gas compression facilities, raised petrochemical tanks, raised piping systems, and roof access points into buildings. In addition, communications equipment, antennae, or other electronic devices may be located on the top of these raised assets. Typically, these assets are reached by ladders that are permanently anchored to the asset. For example, raised petrochemical/water tanks typically are accessed by ladders that are bolted to one of the legs of the tank. Controlling access to these raised assets by controlling access to the ladder can increase security at a Dam Sector facility.

A typical ladder access control system consists of some type of cover that is locked or secured over the ladder. The cover can be a casing that surrounds most of the ladder, or a door or shield that covers only part of the ladder. In either case, several rungs of the ladder (the number of rungs depends on the size of the cover) are made inaccessible by the cover, and these rungs can only be accessed by opening or removing the cover. The cover is locked so that only authorized personnel can open or remove it and use the ladder. Ladder access controls are usually installed at several feet above ground level, and they usually extend several feet up the ladder so that they cannot be circumvented by someone accessing the ladder above the control system.

The important features of ladder access control are the size and strength of the cover and its ability to lock or otherwise be secured from unauthorized access.

The covers are constructed from aluminum or some type of steel. This should provide adequate protection from being pierced or cut through. The metals are corrosion resistant so that they will not corrode or become fragile from extreme weather conditions in outdoor applications. The bolts used to install each of these systems are galvanized steel. In addition, the bolts for each cover are installed on the inside of the unit so they cannot be removed from the outside.

LOCKS

A lock is a type of physical security device that can be used to delay or prevent a door, a gate, a window, a manhole, a filing cabinet drawer, or some other physical feature from being opened, moved, or operated. Locks typically operate by connecting two pieces together—such as by connecting a door to a door jamb or a manhole to its casement. Every lock has two modes—engaged (or "locked") and disengaged (or "opened"). When a lock is disengaged, the asset on which the lock is installed can be accessed by anyone, but when the lock is engaged, only access to the locked asset.

Before discussing locks and their applicability it is important to discuss **key control**. Based on our experience, many Dam Sector facilities (and others) have no idea how many keys for various site/equipment locks have been issued to employees over the years. Many facilities simply issue keys to employees at hiring with no accountability for the keys upon the employee's departure. Needless to say this is not good security policy. You can have the best made locks available installed throughout your facilities but if you do not have proper key control, you do not have proper security.

Locks are excellent security features because they have been designed to function in many ways and to work on many different types of assets. Locks can also provide different levels of security depending on how they are designed and implemented. The security provided by a lock is dependent on several factors, including its ability to withstand physical damage (i.e., can it be cut off, broken, or otherwise physically disabled) as well as its requirements for supervision or operation (i.e., combinations may need to be changed frequently so that they are not compromised and the locks remain secure). While there is no single definition of the "security" of a lock, locks are often described as minimum, medium, or maximum security. Minimum security locks are those that can be easily disengaged (or "picked") without the correct key or code, or those that can be disabled easily (such as small padlocks that can be cut with bolt cutters). Higher security locks are more complex and thus are more difficult to pick, or are sturdier and more resistant to physical damage.

Many locks such as many door locks, only need to be unlocked from one side. For example, most door locks need a key to be unlocked only from the outside. A person opens such devices, called single-cylinder locks, from the inside by pushing a button or by turning a knob or handle. Double-cylinder locks require a key to be locked or unlocked from both sides.

MANHOLES

Manholes are commonly found in Dam Sector industrial sites. Manholes are designed to provide access to the underground utilities, meter vaults, petrochemical pumping rooms, and the like, and therefore they are potential entry points to a system. Because many utilities run under other infrastructure (roads, buildings), manholes also provide potential access points to critical infrastructure as well as petrochemical process assets. In addition, because the portion of the system to which manholes provide entry is primarily located underground, access to a system through a manhole increases the chance that an intruder will not be seen. Therefore, protecting manholes can be a critical component of guarding an entire plant site and a surrounding community.

There are multiple methods for protecting manholes, including preventing unauthorized personnel from physically accessing the manhole (through a manhole lock) and detecting attempts at unauthorized access to the manhole.

A **manhole intrusion sensor** is a physical security device designed to detect unauthorized access to the facility through a manhole. Monitoring a manhole that provides access to a facility or water system can mitigate two distinct types of threats. First, monitoring a manhole may detect access of unauthorized personnel to assets through the manhole. Second, monitoring manholes may also allow the detection of intruders attempting to place explosives or other destructive devices into the system.

Several different technologies have been used to develop manhole intrusion sensors, including mechanical systems, magnetic systems, and fiber optic and infrared sensors. Some of these intrusion sensors have been specifically designed for manholes, while others consist of standard, off-the-shelf intrusion sensors that have been implemented in a system specifically designed for application in a manhole.

RADIATION DETECTION EQUIPMENT FOR MONITORING PERSONNEL AND PACKAGES

One of the major potential threats facing any critical production facility or system is contamination by radioactive substances. Radioactive substances brought on-site at a facility could be used to contaminate the facility, thereby preventing workers from safely entering the facility to perform necessary Dam Sector operations. In addition, radioactive substances brought on-site at a Dam Sector entity could be discharged into the waterways, contaminating the downstream water supply. Therefore, detection of radioactive substances being brought on-site can be an important security enhancement.

Different radionuclides have unique properties, and different equipment is required to detect different types of radiation. However, it is impractical and potentially unnecessary to monitor for specific radionuclides being brought on-site. Instead, for security purposes, it may be more useful to monitor for gross radiation as an indicator of unsafe substances.

In order to protect against these radioactive materials being brought on-site, a facility may set up monitoring sites outfitted with radiation detection instrumentation at entrances to the facility. Depending on the specific types of equipment chosen, this equipment would detect radiation emitted from people, packages, or other objects being brought through an entrance.

One of the primary differences between the different types of detection equipment is the means by which the equipment reads the radiation. Radiation may either be detected by direct measurement or through sampling.

Direct radiation measurement involves measuring radiation through an external probe on the detection instrumentation. Some direct measurement equipment detects radiation emitted into the air around the monitored object. Because this equipment detects radiation in the air, it does not require that the monitoring equipment make physical contact with the monitored object. Direct means for detecting radiation include using a walk-through portal-type monitor that would detect elevated radiation levels on a person or in a package, or by using a handheld detector, which would be moved or swept over individual objects to locate a radioactive source.

Some types of radiation, such as alpha or low energy beta radiation, have a short range and are easily shielded by various materials. These types of radiation cannot be measured through direct measurement. Instead, they must be measured through sampling. Sampling involves wiping the surface to be tested with a special filter cloth, and then reading the cloth in a special counter. For example, specialized smear counters measure alpha and low energy beta radiation.

SECURITY FOR DOORWAYS—SIDE HINGED DOORS

Doorways are the main access points to a facility or to rooms within a building. They are used on the exterior or in the interior of buildings to provide privacy and security for the areas behind them. Different types of doorway security systems may be installed in different doorways depending on the needs or requirements of the buildings or rooms. For example, exterior doorways tend to have heavier doors to withstand the elements and to provide some security to the entrance of the building. Interior doorways in office areas may have lighter doors that may be primarily designed to provide privacy rather than security. Therefore, these doors may be made of glass or lightweight wood. Doorways in industrial areas may have sturdier doors than do other interior doorways and may be designed to provide protection or security for areas behind the doorway. For example, fireproof doors may be installed in chemical storage areas or in other areas where there is a danger of fire.

Because they are the main entries into a facility or a room, doorways are often prime targets for unauthorized entry into a facility or an asset. Therefore, securing doorways may be a major step in providing security at a facility.

A doorway includes four main components:

- The door, which blocks the entrance. The primary threat to the actual door is breaking or piercing through the door. Therefore, the primary security features of doors are their strength and resistance to various physical threats, such as fire or explosions.
- The door frame, which connects the door to the wall. The primary threat to a door frame is that the door can be pried away from the frame. Therefore, the primary security feature of a door frame is its resistance to prying.
- The hinges, which connect the door to the door frame. The primary threat to door hinges is that they can be removed or broken, which will allow intruders to remove the entire door. Therefore, security hinges are designed to be resistant to breaking. They may also be designed to minimize the threat of removal from the door.
- The lock, which connects the door to the door frame. Use of the lock is controlled through various security features, such as keys, combinations, etc., such that only authorized personnel can open the lock and go through the door. Locks may also incorporate other security features, such as software or other systems to track overall use of the door or to track individuals using the door.

Each of these components is integral in providing security for a doorway, and upgrading the security of only one of these components while leaving the other components unprotected may not increase the overall security of the doorway. For example, many facilities upgrade door locks as a basic step in increasing the security of a facility. However, if the facilities do not also focus on increasing security for the door hinges or the door frame, the door may remain vulnerable to being removed from its frame, thereby defeating the increased security of the door lock.

The primary attribute for the security of a door is its strength. Many security doors are 4-20 gauge hollow metal doors consisting of steel plates over a hollow cavity reinforced with steel stiffeners to give the door extra stiffness and rigidity. This increases resistance to blunt force used to try to penetrate through the door. The space between the stiffeners may be filled with specialized materials to provide fire-, blast-, or bullet-resistance to the door.

The Windows and Doors Manufacturers Association have developed a series of performance attributes for doors. These include:

- Structural Resistance
- Forced Entry Resistance
- Hinge Style Screw Resistance
- Split Resistance
- Hinge Resistance
- Security Rating
- Fire Resistance
- Bullet Resistance
- Blast Resistance

The first five bullets provide information on a door's resistance to standard physical breaking and prying attacks. These tests are used to evaluate the strength of the door and the resistance of the hinges and the frame in a standardized way. For example, the Rack Load Test simulates a prying attack on a corner of the door. A test panel is restrained at one end, and a third corner is supported. Loads are applied and measured at the fourth corner. The Door Impact Test simulates a battering attack on a door and frame using impacts of 200 foot pounds by a steel pendulum. The door must remain fully operable after the test. It should be noted that door glazing is also rated for resistance to shattering. Manufacturers will be able to provide security ratings for these features of a door as well.

Door frames are an integral part of doorway security because they anchor the door to the wall. Door frames are typically constructed from wood or steel, and they are installed such that they extend for several inches over the doorway that has been cut into the wall. For added security, frames can be designed to have varying degrees of overlap with, or wrapping over, the underlying wall. This can make prying the frame from the wall more difficult. A frame formed from a continuous piece of metal (as opposed to a frame constructed from individual metal pieces) will prevent prying between pieces of the frame.

Many security doors can be retrofit into existing frames; however, many security door installations including replacing the door frame as well as the door itself. For example, bullet-resistance per Underwriter's Laboratory (UL) 752 requires resistance of the door and frame assembly, and thus replacing the door only would not meet UL 752 requirements.

VALVE LOCKOUT DEVICES

Valves are utilized as control elements in water flow and chemical supply piping networks. They regulate the flow of both liquids and gases by opening, closing, or obstructing a flow passageway. Valves are typically located where flow control is necessary. They can be located in-line or at pipeline and tank entrance and exit points. They can serve multiple purposes in a process pipe network, including:

- Redirecting and throttling flow
- Preventing backflow
- Shutting off flow to a pipeline or tank (for isolation purposes)
- Releasing pressure
- Draining extraneous liquid from pipelines or tanks
- Introducing chemicals into the process network
- As access points for sampling process water

Valves may be located either above ground or below ground. It is critical to provide protection against valve tampering. For example, tampering with a pressure relief valve could result in a pressure buildup and potential explosion in the piping network. On a larger scale, addition of a contaminant or non-compatible chemical substance to the chemical processing system through an unprotected valve could result in the catastrophic release of that contaminant to the general population.

Different security products are available to protect aboveground versus below ground valves. For example, valve lockout devices can purchased to protect valves and valve controls located aboveground. Vaults containing underground valves can be locked to prevent access to these valves.

As described, a lockout device can be used as a security measure to prevent unauthorized access to above-ground valves located within petrochemical processing systems. Valve lockout devices are locks that are specially designed to fit over valves and valve handles to control their ability to be turned or seated. These devices can be used to lock the valve into the desired position. Once the valve is locked, it cannot be turned unless the locking device is unlocked or removed by an authorized individual.

Various valve lockout options are available for industrial use, including:

- Cable lockouts
- Padlocked chains/cables
- Valve-specific lockouts

Many of these lockout devices are not specifically designed for use in Dam Sector industry (i.e., chains, padlocks), but are available from a local hardware store or manufacturer specializing in safety equipment. Other lockout devices (e.g., valve-specific lockouts or valve box-locks) are more specialized and must be purchased from safety or valve-related equipment vendors.

The three most common types of valves for which lockout devices are available are gate, ball, and butterfly valves. Each is described in more detail below.

- **Gate Valve Lockouts**—these are designed to fit over the operating hand wheel of the gate valve to prevent it from being turned. The lockout is secured in place with a padlock. Two types of gate valve lockouts are available: diameter-specific and adjustable. Diameter-specific lockouts are available for handles ranging from 1 inch to 13 inches in diameter. Adjustable gate valve lockouts can be adjusted to fit any handle ranging from 1 inch to 6 inches, or more, in diameter.
- **Ball Valve Lockouts**—There are several different configurations available to lock out ball valves, all of which are designed to prevent rotation of the valve handle. The three major configurations available are a wedge shape

for 1-inch to 3-inch valves, a lockout that completely covers ⅜-inch to 8-inch ball valve handles, and a universal lockout that can be applied to quarter-turn valves of varying sizes and geometric handle dimensions. All three types of ball valve lockouts can be installed by sliding the lockout device over the ball valve handle and securing it with a padlock.

- **Butterfly Valve Lockouts**—This functions in a similar manner to the ball valve lockout. The polypropylene lockout device is placed over the valve handle and secured with a padlock. This type of lockout has been commonly used in the bottling industry.

A major difference between valve-specific lockout devices and the padlocked chain or cable lockouts discussed earlier is that they do not need to be secured to an anchoring device in the floor or the piping system. In addition, valve-specific lockouts eliminate potential tripping or access hazards that may be caused by chains or cable lockouts applied to valves located near walkways or frequently maintained equipment.

Valve-specific lockout devices are available in a variety of colors, which can be useful in distinguishing different valves. For example, different colored lockouts can be used to distinguish the type of liquid passing through the valve (i.e., treated, untreated, potable, petrochemical), or to identify the party responsible for maintaining the lockout. Implementing a system of different-colored locks on operating valves can increase system security by reducing the likelihood of an operator inadvertently opening the wrong valve and causing a problem in the system.

SECURITY FOR VENTS

Vents are installed in some aboveground Dam Sector storage areas to allow safe venting of off-gases. The specific vent design for any given application will vary depending on the design of the chemical storage vessel. However, every vent consists of an open air connection between the storage container and the outside environment. Although these air exchange vents are an integral part of covered or underground chemical storage containers, they also represent a potential security threat. Improving vent security by making the vents tamper-resistant or by adding other security features, such as security screens or security covers, can enhance the security of the entire petrochemical processing system.

Many municipalities already have specifications for vent security at their local chemical industrial assets. These specifications typically include the following requirements:

- Vent openings are to be angled down or shielded to minimize the entrance of surface and/or rainwater into the vent through the opening.
- Vent designs are to include features to exclude insects, birds, animals, and dust.
- Corrosion-resistant materials are to be used to construct the vents.

VISUAL SURVEILLANCE MONITORING

Visual surveillance is used to detect threats through continuous observation of important or vulnerable areas of an asset. The observations can also be recorded for later review or use (e.g., in court proceedings). Visual surveillance system can be used to monitor various parts of production, distribution, or pumping/compressing systems, including the perimeter of a facility, outlying pumping stations, or entry or access points into specific buildings. These systems are also useful in recording individuals who enter or leave a facility, thereby helping to identify unauthorized access. Images can be transmitted live to a monitoring station, where they can be monitored in real time, or they can be recorded and reviewed later. Many Dam Sector facilities have found that a combination of electronic surveillance and security guards provides an effective means of facility security.

Visual surveillance is provided through a **CCTV** system, in which the capture, transmission, and reception of an image is localized within a closed "circuit." This is different than other broadcast images, such as over-the-air television, which is broadcast over the air to any receiver within range.

At a minimum, a CCTV system consists of:

- One or more cameras
- A monitor for viewing the images
- A system for transmitting the images from the camera to the monitor

Specific attributes and features of camera systems, lenses, and lighting systems are presented in table 9.11.

Table 9.11. **Attributes of Camera, Lenses, and Lighting Systems**

Attribute	*Discussion*
	Camera Systems
Camera Type	Major factors in choosing the correct camera are the resolution of the image required and lighting of the area to be viewed. • **Solid State** (including charge coupled devices, charge priming device, charge injection device, and metal oxide substrate)—these cameras are becoming predominant in the marketplace because of their high resolution and their elimination of problems inherent in tub cameras. • **Thermal**—These cameras are designed for night vision. They require no light and use differences in temperature between objects in the field of view to produce a video image. Resolution is low compared to other cameras, and the technology is currently expensive relative to other technologies. • **Tube**—These cameras can provide high-resolution burnout and must be replaced after 1–2 years. In addition, tube performance can degrade over time. Finally, tube cameras are prone to burn images in the tube replacement.
Resolution (the ability to see fine details)	User must determine the amount of resolution required dependingon the level of detail required for threat determination. A high definition focus with a wide field of vision will give an optimal viewing area.
Field of vision width	Cameras are designed to cover a defined field of vision, which is usually defined in degrees. The wider the field of vision, the more area a camera will be able to monitor.
Type of image produced (color, black and white, thermal)	Color images may allow the identification of distinctive markings, while black and white images may provide sharper contrast. Thermal imaging allows the identification of heat sources (such as human beings or other living creatures) from low light environments; however, thermal images are not effective in identifying specific individuals (i.e., for subsequent legal processes).

(*continued*)

Table 9.11. *Continued*

Attribute	*Discussion*
Pan/Tilt/Zoom (PTZ)	Panning (moving the camera in a horizontal plane), tilting (moving the camera in a vertical plane), and zooming (moving the lens to focus on objects that are at different distances from the camera) allow the camera to follow a moving object. Different systems allow these functions to be controlled manually or automatically. Factors to be considered in PTZ cameras are the degree of coverage for pan and tilt function sand the power of the zoom lens.
	Lenses
Format	Lens format determines the maximum image size to be transmitted.
Focal Length	This is the distance from the lens to the center of the focus. The greater the focal length, the higher the magnification but the narrower the field of vision.
F Number	F number is the ability to gather light. Smaller F numbers may be required for outdoor applications where light cannot be controlled as easily.
Distance and width approximation	The distance and width approximations are used to determine the geometry of the space that can be monitored at the best resolution.
	Lighting Systems
Intensity	Light intensity must be great enough for the camera type to produce sharp images. Light can be generated from natural or artificial sources. Artificial sources can be controlled to produce the amount and distribution of light required for a given camera and lens.
Evenness	Light must be distributed evenly over the field of view so that there are no darker or shadowy areas. If there are lighter versus darker areas, brighter areas may appear washed out (i.e., details cannot be distinguished) while no specific objects can be viewed from darker areas.
Location	Light sources must be located above the camera so that light does not shine directly into the camera.

Source: USEPA, 2005.

COMMUNICATION INTEGRATION

In this section, those devices necessary for communication and integration of the Dam Sector industrial processing operations, such as electronic controllers, two-way radios, and wireless data communications are discussed. Typically, SCADA systems would also be discussed in this section; however, SCADA was discussed in detail in chapter 7.

In regards to security applications, electronic controllers are used to automatically activate equipment (such as lights, surveillance cameras, audible alarms, or locks) when they are triggered. Triggering could be in response to variety of scenarios, including tripping of an alarm or a motion sensor; breaking of a window or a glass door; variation in vibration sensor readings; or simply through input from a timer.

Two-way wireless radios allow two or more users that have their radios tuned to the same frequency to communicate instantaneously with each other without the radios being physically lined together with wires or cables.

Wireless data communications devices are used to enable transmission of data between computer systems and/or between a SCADA server and its sensing devices, without individual components being physically linked together via wires or cables. In industrial petrochemical processing systems, these devices are often used to link remote monitoring stations (i.e., SCADA components) or portable computers (i.e., laptops) to computer networks without using physical wiring connections.

Electronic Controllers

An electronic controller is a piece of electronic equipment that receives incoming electric signals an uses preprogrammed logic to generate electronic output signals based on the incoming signals. While electronic controllers can be implemented for any application that involves inputs and outputs (e.g., control of a piece of machinery in a factor), in a security application, these controllers essentially act as the system's "brain," and can respond to specific security-related inputs with preprogrammed output response. These systems combine the control of electronic circuitry with a logic function such that circuits are opened and closed (and thus equipment is turned on and off) through some preprogrammed logic. The basic principle behind the operation of an electrical controller is that it receives electronic inputs from sensors or any device generating an electrical signal (e.g., electrical signals from motion sensors), and then uses it preprogrammed logic to produce electrical outputs (e.g., these outputs could turn on power to a surveillance camera or to an audible alarm). Thus, these systems automatically generate a preprogrammed, logical response to a preprogrammed input scenario.

The three major types of electronic controllers are timers, electromechanical relays, and PLCs, which are often called "digital relays." Each of these types of controller is discussed in more detail next.

Timers use internal signal/inputs (in contrast to externally—generated inputs) and generate electronic output signals at certain times. More specifically, timers control electric current flow to any application to which they are connected, and can turn the current on or off on a schedule pre-specified by the user. Typical timer range (amount of time that can be programmed to elapse before the timer activates linked equipment) is from 0.2 seconds to 10 hours, although some of the more advanced timers have ranges of up to 60 hours. Timers are useful in fixed applications that don't require frequent schedule changes. For example, a timer can be used to turn on the lights in a room or building at a certain time every day. Timers are usually connected to their own power supply (usually 120–240 V).

In contrast to timers, which have internal triggers based on a regular schedule, **electromechanical relays** and **PLCs** have both external inputs and external outputs. However, PLCs are more flexible and more powerful than are electromechanical relays, and thus this section focuses primarily on PLCs as the predominant technology for security-related electronic control applications.

Electromechanical relays are simple devices that use a magnetic field to control a switch. Voltage applied to the relay's input coil creates a magnetic field, which attracts an internal metal switch. This causes the relay's contacts to touch, closing the switch and completing the electrical circuit. This activates any linked equipment. These types of systems are often used for high voltage applications, such as in some automotive and other manufacturing processes.

Two-Way Radios

Two-way radios, as discussed here, are limited to a direct unit-to-unit radio communication, either via single unit-to-unit transmission and reception, or via multiple handheld units to a base station radio contact and distribution system. Radio frequency spectrum limitations apply to all handheld units, and directed by the FCC. This also distinguishes a handheld unit from a base station or base station unit (such as those used by an amateur (ham) radio operator), which operate under different wave length parameters.

Two-way radios allow a user to contact another user or group of users instantly on the same frequency, and to transmit voice or data without the need for wires. They use "half-duplex" communications; or communication that can be only transmitted or received; it can't transmit and receive simultaneously. In other words, only one person may talk, while other personnel with radio(s) can only listen. To talk, the user depresses the talk button and speaks into the radio. The audio then transmits the voice wirelessly to the receiving radios. When the speaker has finished speaking and the channel has cleared, users on any of the receiving radios can transmit; either to answer the first transmission, or to begin a new conversation. In addition to carrying voice data, many types of wireless radios also allow the transmission of digital data, and these radios may be interfaced with computer networks that can use or track these data. For example, some two-way radios can send information such as global positioning system (GPS) data, or the ID of the radio. Some two-way radios can also send data through a SCADA system.

Wireless radios broadcast these voice or data communications over the airwaves from the transmitter to the receiver. While this can be an advantage in that the signal emanates in all directions and does not need a direct physical connection to be received at the receiver, it can also make the communications vulnerable to being blocked, intercepted, or otherwise altered. However, security features are available to ensure that the communications are not tampered with.

Wireless Data Communications

A wireless data communication system consists of two components: a "Wireless Access Point" (WAP), and a "Wireless Network Interface Card" (sometimes also referred to as a "Client"), which work together to complete the communications link. These wireless systems can link electronic devices, computers, and computer systems together using radio waves, thus eliminating the need for these individual components to be directly connected together through physical wires. While wireless data communications have widespread application in water and wastewater systems, they also have limitations. First, wireless data connections are limited by the distance between components (radio waves scatter over a long distance and cannot be received efficiently, unless special directional antenna are used). Second, these devices only function if the individual components are in a direct line of sight with each other, since radio waves are affected by interference from physical obstructions. However, in some cases, repeater units can be used to amplify and retransmit wireless signals to circumvent these problems. The two components of wireless devices are discussed in more detail here.

(1) **WAP**: The WAP provides the wireless data communication service. It usually consists of a housing (which is constructed from plastic or metal depending on the environment it will be used in) containing a circuit board; flash memory that holds software; one of two external ports to connect to existing wired networks; a wireless radio transmitter/receiver; and one or more antenna connections. Typically, the WAP requires a one-time user configuration to allow the device to interact with the Local Area Network (LAN). This configuration is usually done via a web-driven software application which is accessed via a computer.

(2) **Wireless Network Interface Card/Client**: A wireless card is a piece of hardware that is plugged in to a computer and enables that computer to make a wireless network connection. The card consists of a transmitter, functional circuitry, and a receiver for the wireless signal, all of which work together to enable communication between the computer, its wireless transmitter/receiver, and its antenna connection. Wireless cards are installed in a computer through a variety of connections, including USB Adapters, or Laptop CardBus (PCMCIA) or Desktop Peripheral (PCI) cards. As with the WAP, software is loaded onto the user's computer, allowing configuration of the card so that it may operate over the wireless network.

Two of the primary applications for wireless data communications systems are to enable mobile or remote connections to a LAN, and to establish wireless communications links between SCADA RTUs and sensors in the field. Wireless car connections are usually used for LAN access from mobile computers. Wireless cards can also be incorporated into RTUs to allow them to communicate with sensing devices that are located remotely.

CYBER PROTECTION DEVICES

Various cyber protection devices are currently available for use in protecting Dam Sector computer systems. These protection devices include anti-virus and pest eradication software, firewalls, and network intrusion hardware/software. These products are discussed in this section.

Anti-Virus and Pest Eradication Software

Anti-virus programs are designed to detect, delay, and respond to programs or pieces of code that are specifically designed to harm computers. These programs are known as "malware." Malware can include computer viruses, worms, and Trojan Horse programs (programs that appear to be benign but which have hidden harmful effects).

Pest eradication tools are designed to detect, delay, and respond to "spyware" (strategies that websites use to track user behavior, such as by sending "cookies" to the user's computer), and hacker tools that track keystrokes (keystroke loggers) or passwords (password crackers).

Viruses and pests can enter a computer system through the Internet or through infected floppy discs or CDs. They can also be placed onto a system by insiders. Some of these programs, such as viruses and worms, move within a computer's drives and files, or between computers if the computers are networked to each other. This malware can deliberately damage files, utilize memory and network capacity, crash application programs, and initiate transmissions of sensitive information from a PC. While the specific mechanisms of these programs differ, they can infect files, and even the basic operating program of the computer firmware/hardware.

The most important features of an anti-virus program are its abilities to identify potential malware and to alert a user before infection occurs, as well as its ability to respond to a virus already resident on a system. Most of these programs provide a log so that the user can see what viruses have been detected and where they were detected. After detecting a virus, the anti-virus software may delete the virus automatically, or it may prompt the user to delete the virus. Some programs will also fix files or programs damaged by the virus.

Various sources of information are available to inform the general public and computer system operators about new viruses being detected. Since anti-virus programs use signatures (or snippets of code or data) to detect the presence of a virus, periodic updates are required to identify new threats. Many anti-virus software providers offer free upgrades that are able to detect and respond to the latest viruses.

Firewalls

A firewall is an electronic barrier designed to keep computer hackers, intruders or insiders from accessing specific data files and information on a Dam Sector's computer network or other electronic/computer systems. Firewalls operated by evaluating and then filtering information coming through a public network (such as the Internet) into the utility's computer or other electronic system. This evaluation can include identifying the source or destination addresses and ports, and allowing or denying access based on this identification.

There are two methods used by firewalls to limit access to the utility's computers or other electronic systems from the public network:

- The firewall may deny all traffic unless it meets certain criteria.
- The firewall may allow all traffic through unless it meets certain criteria.

A simple example of the first method is to screen requests to ensure that they come from an acceptable (i.e., previously identified) domain name and Internet protocol address. Firewalls may also use more complex rules that analyze the application data to determine if the traffic should be allowed through. For example, the firewall may require user authentication (i.e., use of a password) to access the system. How a firewall determines what traffic to let through depends on which network layer it operates at and how it is configured. Some of the pros and cons of various methods to control traffic flowing in and out of the network are provided in table 9.12.

Firewalls may be a piece of hardware, a software program, or an appliance card that contains both.

Advanced features that can be incorporated into firewalls allow for the tracking of attempts to log-on to the LAN system. For example, a report of successful and unsuccessful log-in attempts may be generated for the computer specialist to analyze. For systems with mobile users, firewalls allow remote access in to the private network by the use of secure log-on procedures

Table 9.12. Pros and Con of Various Firewall Methods for Controlling Network Access

Method	*Description*	*Pros*	*Cons*
Packet Filtering	Incoming and outgoing packets (small chunks of data) are analyzed against a set of filters. Packets that make it through the filters are sent to the requesting system and all others are discarded. There are two type of packet filtering: static (the most common) and dynamic.	Static filtering is relatively inexpensive, and little maintenance required. It is well suited for closed environments where access to or from multiple addresses is not allowed.	Leaves permanent open holes in the network; allows direct connection to internal hosts by external sources; offers no user authentication, method can be unmanageable in large networks.
Proxy Service	Information from the Internet is retrieved by the firewall and then sent to the requesting system and vice versa. In this way, the firewall can limit the information made known to the requesting system, making vulnerabilities less apparent.	Only allows temporary open holes in the network perimeter. Can be used for all types of internal protocol services.	Allows direct connections to internal hosts by external clients; offers no user authentication.
Stateful Pattern Recognition	This method examines and compares the contents of certain key parts of an information packet against a database of acceptable information. Information traveling from inside the firewall to the outside is monitored for specific defining characteristics, then incoming information is compared to these characteristics. If the comparison yield a reasonable match, the information is allowed through. If not, the information is discarded.	Provides a limited time window to allow pockets of information to be sent; does not allow any direct connections between internal and external hosts; supports user-level authentication.	Slower than packet filtering; does not support all types of connections.

Source: USEPA, 2005.

and authentication certificates. Most firewalls have a graphical user interface for managing the firewall.

In addition, new Ethernet firewall cards that fit in the slot of an individual computer bundle additional layers of defense (like encryption and permit/deny) for individual computer transmissions to the network interface function. These new cards have only a slightly higher cost than traditional network interface cards.

Network Intrusion Hardware/Software

Network intrusion detection and prevention system are software- and hardware-based programs designed to detect unauthorized attacks on a computer network system.

While other applications, such as firewalls and anti-virus software, share similar objectives with network intrusion systems, network intrusion systems provide a deeper layer of protection beyond the capabilities of these other systems because they evaluate pattern of computer activity rather than specific files.

It is worth noting that attacks may come from either outside or within the system (i.e., from an insider), and that network intrusion detection systems may be more applicable for detecting patterns of suspicious activity from inside a facility (i.e., accessing sensitive data) than are other information technology solutions.

Network intrusion detection systems employ a variety of mechanisms to evaluate potential threats. The types of search and detection mechanisms are dependent upon the level of sophistication of the system. Some of the available detection methods include:

- **Protocol analysis**—Protocol analysis is the process of capturing, decoding, and interpreting electronic traffic. The protocol analysis method of network intrusion detection involves the analysis of data captured during transactions between two or more systems or devices, and the evaluation of these data to identify unusual activity and potential problems. Once a problem is isolated and recorded, problems or potential threats can be linked to pieces of hardware or software. Sophisticated protocol analysis will also provide statistics and trend information on the captured traffic.

- **Traffic anomaly detection**—Traffic anomaly detection identifies potential threatening activity by comparing incoming traffic to "normal" traffic patterns, and identifying deviations. It does this by comparing user characteristics against thresholds and triggers defined by the network administrator. This method is designed to detect attacks that span a number of connections, rather than a single session.
- **Network honeypot**—This method establishes nonexistent services in order to identify potential hackers. A network honeypot impersonates services that don't exist by sending fake information to people scanning the network. It identifies the attacker when they attempt to connect to the service. There is no reason for legitimate traffic to access these resources because they don't exist; therefore any attempt to access them constitutes an attack.
- **Anti-intrusion detection system evasion techniques**—These methods are designed attackers who may be trying to evade intrusion detection system scanning. They include methods called IP defragmentation, TCP streams reassembly, and deobfuscation.

While these detection systems are automated, they can only indicate patterns of activity and a computer administer or other experienced individual must interpret activities to determine whether or not they are potentially harmful. Monitoring the logs generated by these systems can be time consuming, and there may be a learning curve to determine a baseline of "normal" traffic patterns from which to distinguish potential suspicious activity.

REFERENCES AND RECOMMENDED READING

Garcia, M. L. 2001. *The Design and Evaluation of Physical Protection Systems.* Butterworth-Heinemann.

IBWA. 2004. *Bottled Water Safety and Security*. Alexandria, VA: International Bottled Water Association.

NAERC. 2002. *Security Guidelines for the Electricity Sector*. Washington, DC: North American Electric Reliability Council.

USEPA. 2005. *Water and Wastewater Security Product Guide.* http://cfpub.epa.gov .safewater/watersecurity/guide (accessed April 4, 2009).

Appendix

Definition of Terms

Acid bomb—A crude bomb made by combining muriatic acid with aluminum strips in a 2-liter soda bottle.
Acre-foot—The volume of water that would cover 1 acre to a depth of 1 foot.
Aerosol—A fine mist or spray, which contains minute particles.
Aflatoxin—A toxin created by bacteria that grow on stored foods, especially on rice, peanuts, and cotton seeds.
Agency—A division of government with a specific function, or a nongovernmental organization (e.g., private contractor, business) that offers a particular kind of assistance. In the incident command system, agencies are defined as jurisdictional (having statutory responsibility for incident mitigation) or assisting and/or cooperating (providing resources and/or assistance).
Airborne—Carried by or through the air.
al-Gama'a al-Islamiyya (The Islamic Group, IG)—Islamic terrorist group that emerged spontaneously during the 1970s in Egyptian jails and later in Egyptian universities. After President Sadat released most of the Islamic prisoners form prisons in 1971, groups of militants organized themselves in groups and cells, and al-Gama'a al-Islamiyya was one of them.
Al-Qaeda—Meaning "the Base"; an international terrorist group founded in approximately 1989 and dedicated to opposing non-Islamic governments with force and violence. One of the principal goals of Al-Qaeda was to drive the U.S-armed forces out of the Saudi Arabian peninsula and Somalia by

violence. Currently wanted for several terrorist attacks, including those on the U.S. embassy in Kenya and Tanzania as well as the first and second World Trade Center bombings, and the attack on the Pentagon.

al Tahwid—A Palestinian group based in London which professes a desire to destroy both Israel and the Jewish people throughout Europe. Eleven al Tahwid were arrested in Germany allegedly as they were about to begin attacking that country.

All-Hazards—A grouping classification encompassing all conditions, environmental or manmade, that have the potential to cause injury, illness, or death; damage to or loss of equipment, infrastructure services, or property, or alternatively causing functional degradation to social, economic, or environmental aspects.

Alpha radiation—The least penetrating type of nuclear radiation. Not considered dangerous unless particles enter the body.

Alternating current—An electric current changing regularly (alternating) from one direction to the opposite.

Ammonium nitrate-fuel oil—A powerful explosive made by mixing fertilizer and fuel oil. The type of bomb used in the first World Trade Center attack as well as Oklahoma City bombing.

Ampere—The common unit of measurement of electrical current.

Analyte—The name assigned to a substance or feature that describes it in terms of its molecular composition, taxonomic nomenclature, or other characteristic.

Anthrax—An often fatal infectious disease contracted from animals. Anthrax spores have a long survival period; the incubation period is short; disability is severe—making anthrax a bioweapon of choice by several nations.

Antidote—A remedy to counteract the effects of poison.

Antigen—A substance which stimulates an immune response by the body immune system recognizes such substances as foreign and produces antibodies to fight them.

Antitoxin—An antibody which neutralizes a biological toxin.

Asset—A person, structure, facility, information, material or process that has value. In the context of the National Infrastructure Protection Plan (NIPP), people are not considered assets.

Asymmetric threat—The use of crude or low-tech methods to attack a superior or more high-tech enemy.

Baseload—The minimum constant amount of load connected to the power system over a given time period, usually on a monthly, seasonal, or yearly basis.
Baseload plant—A plant, usually housing high-efficiency steam-electric units, which is normally operated to take all or part of the minimum load of a system, and which consequently produces electricity at an essentially constant rate and runs continuously. These units are operated to maximize system mechanical and thermal efficiency and minimize system operating costs.
Bioaccumulative—Substances that concentrate in living organisms as the breather contaminated air drink or live in contaminated water or eat contaminated food rather than being eliminated through natural processes.
Biochemical warfare—Collective term for use of both chemical warfare and biological warfare weapons.
Biochemterroism—Terrorism using weapons biological or chemical agents.
Biological ammunition—Ammunition designed specifically to release a biological agent used as the warhead for biological weapons. Biological ammunition may take many forms, such as a missile warhead or bomb.
Biological attacks—The deliberate release of germs or other biological substances that cause illness.
Biometric security systems—Systems that use biometric technology (measuring the unique physical characteristics or traits of the human body) combined with some type of locking mechanisms to control access to specific assets.
Biosafety Level 1—Suitable for work involving well-characterized biological agents not known to consistently cause disease in healthy adult humans, and of minimal potential hazard to lab personnel and the environment. Work is generally conducted on open bench tops using standard microbiological practices.
Biosafety Level 2—Suitable for work involving biological agents of moderate potential hazard to personnel and the environment. Lab personnel should have specific training in handling pathogenic agents and be directed by competent scientists. Access to the lab should be limited when work is being conducted, extreme precautions should be taken with contaminated sharp items, and certain procedures should be conducted in biological safety cabinets or other physical containment equipment if there is a risk of creating infectious aerosols or splashes.
Biosafety Level 3—Suitable for work done with indigenous or exotic biological agents that may cause serious or potentially lethal disease as a result of

exposure by inhalation. Lab personnel must have specific training in handling pathogenic and potentially lethal agents and be supervised by competent scientists who are experienced in working with these agents. All procedures involving the manipulation of infectious material are conducted within biological safety cabinets or other physical containment devices, or by personnel wearing appropriate personal protective clothing and equipment. The lab must have special engineering and design features.

Biosafety Level 4—Suitable for work with the most infectious biological agents. Access to the two Biosafety Level 4 labs in the United States is highly restricted.

Bioterrorism—The use of biological agents in a terrorist operation. Biological toxin would include anthrax, ricin, botulism, the plague, smallpox, and tularemia.

Bioterrorism Act—The Public Health Security and Bioterrorism Preparedness and Response Act of 2002.

Biowarfare—The use of biological agents to cause harm to targeted people either directly, by bringing the people into contact with the agents or indirectly, by infecting other animals and plants, which would in turn cause harm to the people.

"Black Start" capabilities—Incremental hydropower available at sites with existing hydroelectric facilities. Incremental hydropower is defined as capacity additions or improved efficiency at existing hydro projects.

Blister agents—Agents which cause pain and incapacitation instead of death and might be used to injure many people at once, thereby overloading medical facilities and causing fear in the population. Mustard gas is the best known blister agent.

Blood agents—Agents based on cyanide compounds. More likely to be used for assassination than for terrorism.

Botulism—An illness caused by the botulinum toxin, which is exceedingly lethal and quite simple to produce. It takes just a small amount of the toxin to destroy the central nervous system. Botulism may be contracted by the ingestion of contaminated food or through breaks or cuts in the skin. Food supply contamination or aerosol dissemination of the botulinum toxin are the two ways most likely to be used by terrorists.

Bus (Buswork)—A conductor, or group of conductors, that serve as a common connection for two or more electrical circuits. In power plants, buswork comprises the three rigid single-phase connectors that interconnect the generator and the step-up transformer(s).

Bush Doctrine—The policy that holds responsible nations which harbor or support terrorist organizations and says that such countries are considered hostile to the United States. From President Bush's speech: "A country that harbors terrorists will either deliver the terrorist or share in their fate. . . . People have to choose sides. They are either with the terrorists, or they're with us."

Business continuity—The ability of an organization to continue to function before, during, and after a disaster.

BWC—Officially known as the "Convention on the Prohibition of Development, Production, and Stockpiling of Bacteriological (Biological) and Toxin Weapons and Destruction." The BWC works toward general and complete disarmament, including the prohibition and elimination of all types of weapons of mass destruction.

Camp X-Ray—The Guantanamo Bay, Cuba prison, which houses Al-Qaeda and Taliban prisoners.

Capability—The maximum load that a generating unit, generating station, or other electrical apparatus can carry under specified conditions for a given period of time without exceeding approved limits of temperature and stress.

Capacity—The amount of electric power delivered or required for which a generator, turbine, transformer, transmission circuit, station, or system is rated by the manufacturer.

Carrier—A person or animal that is potentially a source of infection by carrying on infectious agent without visible symptoms of the disease.

Cascading event—The occurrence of one event that causes another event.

Causative agent—The pathogen, chemical, or other substance that is the cause of disease or death in an individual.

Cell—The smallest unit within a guerrilla or terrorist group. A cell generally consists of two to five people dedicated to a terrorist cause. The formation of cells is born of the concept that an apparent "leaderless resistance" makes it hard for counterterrorists to penetrate.

Chain of custody—The tracking and documentation of physical control of evidence.

Chemical agent—A toxic substance intended to be used for operations to debilitate, immobilize, or kill military or civilian personnel.

Chemical ammunition—A munition, commonly a missile, bomb, rocket, or artillery shell, designed to deliver chemical agents.
Chemical attack—The intentional release of toxic liquid, gas or solid in order to poison the environment or people.
Chemical warfare—The use of toxic chemicals as weapons, not including herbicide used to defoliate battlegrounds or riot control agents such as gas or mace.
Chemical weapons—Weapons that produce effects on living targets via toxic chemical properties. Examples would be sarin, VX nerve gas, or mustard gas.
Chemterrorism—The use of chemical agents in a terrorist operation. Well-known chemical agents include sarin and VX nerve gas.
Choking agent—Compounds that injure primarily in the respiratory tract (i.e., nose, throat, and lungs). In extreme cases membranes swell up, lungs become filled with liquid, and death results from lack of oxygen.
Circuit—A conductor or a system of conductors through which electric current flows.
Cipro—A Bayer antibiotic that combats inhalation anthrax.
"Confirmed"—In the context of the threat evaluation process, a water contamination incident is definitive evidence that the water has been contaminated.
Consequence—The effect of an event, incident, or occurrence. For the purposes of the NIPP, consequences are divided into four main categories: public health and safety, economic, psychological, and governance impacts.
Consequence analysis—The estimate of the potential public health and economic impacts that a successful attack could cause.
Control center—A sophisticated monitoring and control system responsible for balancing power generation and demand; monitoring flows over transmission lines to avoid overloading; planning and configuring the system to operate reliably; maintaining system stability; preparing for emergencies; and placing equipment in and out of service for maintenance and emergencies.
Control systems—Computer-based systems used within many infrastructures and industries to monitor and control sensitive processes and physical functions. These systems typically collect measurement and operation data from the field, process and display the information, and relay control commands to local or remote equipment or human-machine interfaces (operators). Examples of types of control systems include SCADA system, Process Control Systems, and Distributed Control Systems.

Counterterrorism—Measures used to prevent preempt, or retaliate against terrorist attacks.

"Credible"—In the context of the threat evaluation process, a water contamination threat is characterized as "credible" if information collected during the threat evaluation process corroborates information from the threat warning.

Critical infrastructure—Systems and assets, whether physical or virtual, so vital that the incapacity or destruction of such may have a debilitating impact on the security, economy, public health or safety, environment, or any combination of these matters, across any federal, state, regional, territorial, or local jurisdiction.

Criticality—A description of the importance of a particular sector asset, system, network, or function in relation to national or regional security issues. Includes a consideration of public health and economic impacts.

Current (Electric)—A flow of electrons in an electrical conductor. The strength or rate of movement of the electricity is measured in amperes.

Cutaneous—Related to or entering through the skin.

Cutaneous anthrax—Anthrax that is contracted via broken skin. The infection spreads through the bloodstream causing cyanosis, shock, sweating, and finally death.

Cyanide agent—Used by Iraq in the Iran war against the Kurds in the 1980s, and also by the Nazis in the gas chambers of concentration camps, cyanide agents are colorless liquid which is inhaled in its gaseous form while liquid cyanide and cyanide salts are absorbed by the skin. Symptoms are headache, palpitations, dizziness, and respiratory problems followed later by vomiting, convulsions, respiratory failure and unconsciousness and eventually by death.

Cybersecurity—The prevention of damage to, unauthorized use of, or exploitation of, and, if needed, the restoration of electronic information and communications systems and the information contained therein to ensure confidentially, integrity, and availability. Includes protection and restoration, when needed, or information networks and wireline, wireless, satellite, public safety answering points, and 911 communications systems and control systems.

Cyber system—Any combination of facilities, equipment, personnel, procedures, and communications integrated to provide cyber services. Examples include business systems, control systems, and access control systems.

Cyberterrorism—Attacks on computer networks or systems, generally by hackers working with or for terrorist groups. Some forms of cyberterrorism include denial of service attacks, inserting viruses or stealing data.

Dams—Facilities that include some of all of the following components to perform their intended purposes safely: conventional dam or retaining structure (the structural sections that hold back the water), reservoir (the body of water impounded by the dam), spillways (the structure that passes normal and/or flood flows in a manner that protects the structural integrity of the dam), outlet works (the combination of structures and equipment required for safe operation and control of water released from a reservoir), powerhouse (the structure that houses turbines, generators, and associated control equipment for the production of hydroelectricity), and canal/aqueduct (a constructed channel, usually open, that conveys water by gravity or is used for navigation).

Demand—The rate at which electric energy is delivered to or by a system, part of a system, or a piece of equipment. It is expressed in kilowatts, kilovolt amperes, or other suitable units at a given instant or averaged over any designated period of time. The primary source of "demand" is the power-consuming equipment of the customers.

Dependency—The one-directional reliance of an asset, system network, or collection thereof, within or across sectors, in input, interaction, or other requirement from other sources in order to function properly.

Direct current—Electric current going in one direction only.

Dirty bomb—A makeshift nuclear device which is created from radioactive nuclear waste material. While not a nuclear blast, an explosion of a dirty bomb causes localized radioactive contamination as the nuclear waste material is carried into the atmosphere where it is dispersed by the wind.

Distribution system—The portion of an electric system that is dedicated to delivering electric energy to an end user. The distribution system "steps down" power (using step-down transformers) from high-voltage transmission lines to a level that can be used in homes and businesses.

Ebola—Ebola hemorrhagic fever (Ebola EF) is a severe, often-fatal disease in nonhuman primates such as monkeys, chimpanzees, and gorillas and in humans. Ebola has appeared sporadically since 1976 when it was first recognized.

eBomb (for e-bomb)—Electromagnetic bomb which produces a brief pulse of energy which affects electronic circuitry. At low levels, the pulse temporarily disables electronics systems, including computers, radios, and transportation systems. High levels completely destroy circuitry, causing mass disruption of infrastructure while sparing life and property.

Ecotage—Is the portmanteau of the "eco-" prefix and "sabotage." It is used to describe illegal acts of vandalism and violence, committed in the name of environmental protection.

Ecoterrorism—A neologism for terrorism that includes sabotage intended to hinder activities that are considered damaging to the environment.

Electromagnetic pulse—A burst of electromagnetic radiation by deliberate means, such as nuclear attack, or through natural means, such as a large-scale geomagnetic storm. Magnetic and electric fields resulting from EMP have the potential to disrupt electrical and electronic systems by causing destructive current and voltage surges.

Energy—The capacity for doing work as measured by the capability of doing work (potential energy) or the conversion of this capability to motion (kinetic energy). Energy has several forms, some of which are easily convertible and can be changed to another form useful for work. Most of the world's convertible energy comes from fossil fuels that are burned to produce heat that is then used as a transfer medium to mechanical or other means in order to accomplish tasks. Electrical energy is usually measured in kilowatt hours and represents power (kilowatts) operating for some time period (hours), while heat energy is usually measured in British Thermal Units (BTUs).

Energy Asset and System Parameters—Six general asset or system characteristics that are important parameters for evaluating the vulnerabilities of energy infrastructure and developing risk management programs. They include physical and location attributes, cyber attributes, volumetric or throughput attributes, temporal/load profile attributes, human attributes, and the importance of an asset of system to the energy network.

Euroterrorism—Associated with left-wing terrorism of the 1960s, 1970s, and 1980s involving the Red Brigade, Red Army Faction, and November 17th Group, among other groups which targeted American interests in Europe and NATO. Other groups include Orange Volunteers, Red Hand Defenders, Continuity IRA, Loyalist Volunteer Force, Ulster Defense Association, and First of October Anti-Fascist Resistance Group.

Fallout—The descent to the earth's surface of particles contaminated with radioactive material from a radioactive cloud. The term can also be applied to the contaminated particulate matter itself.

Fatah—Meaning "conquest by means of jihad"; a political organization created in the 1960s and led by Yasser Arafat. With both a military and intelligence wing, it has carried out terrorist attacks on Israel since 1965. It joined the PLO in 1968. Since 9/11, the Fatah was blamed for attempting to smuggle 50 tons of weapons into Israel.

Fatwa—A legal ruling regarding Islamic Law.

Filtrate—In ultrafiltration, the water that passes through the membrane and which contains particles smaller than the molecular weight cutoff of the membrane.

Frustration-aggression hypothesis—A hypothesis that every frustration leads to some form of aggression and every aggressive act results from some prior frustration. As defined by Gurr: "The necessary precondition for violent civil conflict is relative deprivation, defined as actors' perception of discrepancy between their value expectations and their environment's apparent value capabilities. This deprivation may be individual or collective."

Function—A service, process, capability, or operation performed by an asset, system, network, or organization.

Fundamentalism—Conservative religious authoritarianism. Fundamentalism is not specific to Islam; it exists in all faiths. Characteristics include literal interpretation of scriptures and a strict adherence to traditional doctrines and practices.

Generation (Electricity)—The process of producing electric energy by transforming other forms of energy; also, the amount of electric energy produced, express in watt-hours (Wh).

Generator—A machine that converts mechanical energy into electrical energy.

Geneva Protocol 1925—The first treaty to prohibit the use of biological weapons. The 1925 Geneva Protocol for the Prohibition of the Use In War of Asphyxiating, Poisonous or Other Gases and Bacteriological Methods of Warfare.

Germ warfare—The use of biological agents to cause harm to targeted people either directly, by bringing the people into contact with the agents or indirectly, by infecting other animals and plants, which would in turn cause harm to the people.

Glanders—An infectious bacterial disease known to cause inflammation in horses, donkeys, mules, goats, dogs and cats. Human infection has not been seen since 1945, but because so few organisms are required to cause disease, it is considered a potential agent for biological warfare.

Government Coordinating Council—The government counterpart to the SCC of each sector established to enable interagency coordination. The GCC comprises representatives across various levels of government (Federal, State, local, tribal, and territorial) as appropriate to the security and operations landscape of each individual sector.

Grab sample—A single sample collected at a particular time and place that represents the composition of the water, air, or soil only at that time and location.

Ground zero—From 1946 until 9/11, ground zero was the point directly above, below, or at which a nuclear explosion occurs or the center or origin of rapid, intense, or violent activity or change. After 9/11, the term, when used with initial capital letters, refers to the ground at the epicenter of the World Trade Center attacks.

Guerrilla warfare—The term was invented to describe the tactics Spain used to resist Napoleon, though the tactic itself has been around much longer. Literally, it means "little war." Guerilla warfare features cells and utilizes no front line. The oldest form of asymmetric warfare, guerilla warfare is based on sabotage and ambush with the objective of destabilizing the government through lengthy and low-intensity confrontation.

Hacktivist—One who hacks into computer networks to achieve a certain political aim—sometimes only to create disruption or to direct attention to a specific cause.

Hamas—A radical Islamic organization which operates primarily in the West Bank and Gaza Strip whose goal is to establish an Islamic Palestinian state in place of Israel. On the one hand, Hamas operates overtly in their capacity as social services deliverers, but its activists have also conducted many attacks, including suicide bombings, against Israeli civilians and military targets.

Hazard—An inherent physical or chemical characteristic that has the potential for causing harm to people, the environment, or property; a natural or man-made source of cause of harm or difficulty.

Hazard assessment—The process of evaluating available information about the site to identify potential hazards that might pose a risk to the site characterization team. The hazard assessment results in assigning one of four levels to risk: lower hazard, radiological hazard, high chemical hazard, or high biological hazard.

Head—The difference in elevation between the headwater surface above and the tailwater surface below a hydroelectric power plant under specified conditions.

Hemorrhagic fevers—In general, the term viral hemorrhagic fever is used to describe severe multisystem syndrome wherein the overall vascular system is damaged, and the body becomes unable to regulate itself. These symptoms are often accompanied by hemorrhage; however, the bleeding itself is not usually life-threatening. While some types of hemorrhagic fever viruses can cause relatively mild illnesses.

High-Impact, Low-Frequency (HILF)—HILF evens are occurrences that are relatively unusual, but have the potential to cause catastrophic disruption. Examples include pandemic disease, terrorist attack, and electromagnet pulse.

Hizbollah (Hezbollah)—Meaning "The Party of God." One of many terrorist organizations which seek the destruction of Israel and of the United States. They have taken credit for numerous bombings against civilians, and have declared that civilian targets are warranted. Hezbollah claims it sees no legitimacy for the existence of Israel, and that their conflict becomes one of legitimacy that is based on religious ideals.

Homeland Security Office—An agency organized after 9/11, originally headed by former Pennsylvania Governor Tom Ridge. The Office of Homeland Security is at the top of approximately 40 federal agencies charged with protecting the United States against terrorism.

Homicide bombings—A term the White House coined the term to replace the old "suicide bombings."

Horsepower—A unit of rate of doing work equal to 550 foot pounds per minute or 745.8 watts (Brit.), 746 watts (USA), or 736 watts (Europe).

Hydrant lock—A physical security device designed to prevent unauthorized access to the water supply through a hydrant. They can also ensue water and water pressure availability to fire fighters and prevent water theft and associated lost water revenue. These locks are usually steel covers or caps that are locked in place over the operating nut of a fire hydrant.

Hydroelectric power—Electric current produced from water power.

Hydroelectric powerplant—A building in which turbines are operated, to drive generators, by the energy of natural or artificial waterfalls.

Incident—A confirmed occurrence that requires response actions to prevent or minimize loss of life or damage to property and/or natural resources. A drinking water contamination incident occurs when the presence of a harmful contaminant has been confirmed.

Infrastructure—The framework of interdependent networks and systems comprising identifiable industries, (including people and procedures), and distribution capabilities that provide a reliable flow of products and services essential to the defense of economic security of the United States, the smooth functioning of government at all levels, and society as a whole. Consistent with the definition in the Homeland Security Act, infrastructure includes physical, cyber, and/or human elements.

Inhalation anthrax—A form of anthrax that is contracted by inhaling anthrax spores. This results in pneumonia, sometimes meningitis, and finally death.

Islamic State of Iraq and the Levant—Islamic extremist rebel group ISIS that presently controls territory and brutalizes and/or murders various groups in various countries, including Syria, Iraq, Libya, Nigeria, and others.

Interdependency—Mutually reliant relationship between entities (objects, individuals, or groups). The degree of interdependency does not need to be equal in both directions.

Intifada (alternatively **Intifadah**, from Arabic "shaking off")—The two intifadas are similar in that both were originally characterized by civil disobedience by the Palestinians which escalated into the use of terror. In 1987, following the killing of several Arabs in the Gaza Strip, the first intifada began and went on until 1993. The second intifada began in September 2000, following Ariel Sharon's visit to the Temple Mount.

Jihad—Meaning "struggle." The definition is a subject of vast debate. There are two definitions generally accepted. The first is a struggle against oppression, whether political or religious. The second is the struggle within oneself, or a spiritual struggle.

Key resources—As defined in the Homeland Security Act, key resources are publicly or privately controlled resources essential to the minimal operations of the economy and government.

Kilowatt (kW)—Unit of electric power equal to 1,000 watts or about 1.34 horsepower. For example, it's the amount of electric energy required to light then 1000-watt light bulbs.

Kilowatt-hour (kWh)—The unit of electrical energy commonly used in marketing electric power; the energy produced by 1 kilowatt acting for one hour. Ten 100-watt light bulbs burning for an hour would consume one kilowatt hour of electricity.

Kinetic energy—Energy which a moving body has because of its motion, dependent on its mass and the rate at which it is moving.

LD50—The dose of a substance which kills 50 percent of those infected.

Laboratory Response Network—A network of labs developed by the CDC, APHL, and FBI for the express purpose of dealing with bioterrorism threats, including pathogens and some biotoxins.

Lassa fever—An acute, often fatal, viral disease characterized by high fever, ulcers of the mucous membranes, headaches, and disturbances of the gastrointestinal system.

Levee—Man-made flood protection systems (embankment sections, floodwalls, closure structures, pumps, interior drainage works, and flood damage reduction channels) with the primary purpose of furnishing protection from seasonal high water (floods), storm surges, precipitation, and other weather events.

Lifeline—Important civil works infrastructures such as bridges, pipelines, buildings, and so forth.

Links—The means (road, rail, barge, or pipeline) by which a chemical is transported from one node to another.

Load (electric)—The amount of electric power delivered or required at any specific point or points on a system. The requirement originates at the energy-consuming equipment of the consumers.

Megawatt—A unit of power equal to 1 million watts. For example, it's the amount of electric energy required to light 10,000 100-watt bulbs.

Mindset—According to *American Heritage Dictionary*: "1. A fixed mental attitude or disposition that predetermines a person's response to an interpretation of situations; 2. and inclination or a habit." *Merriam Webster's Collegiate Dictionary* (10th ed.) defines it as 1. A mental attitude or inclination; 2. a fixed state of mind. The term dates from 1926 but apparently is not included in dictionaries of psychology.

Mitigation—Ongoing and sustained action to reduce the probability of or lessen the impact of an adverse incident.

Molotov cocktail—A crude incendiary bomb made of a bottle filled with flammable liquid and fitted with a rag wick.

Monkeypox—The Russian bioweapon program worked with this virus, which is in the same family as smallpox. In June 2003, a spate of human monkeypox cases was reported in the U.S. Midwest. This was the first time that monkeypox was seen in North America, and it was the first time that monkeypox was transferred from animal to human. There was some speculation that it was a bioattack.

Mustard gas—Blistering agents which cause severe damage to the eyes, internal organs, and respiratory system. Produced for the first time in 1822, mustard gas was not used until World War I. Victims suffered the effects of mustard gas 30 to 40 years after exposure.

Narcoterrorism—The view of many counterterrorist experts that there exists an alliance between drug traffickers and political terrorists.

National Pharmaceutical Stockpile—A stock of vaccines and antidotes which are stored at Centers for Disease Control and Prevention in Atlanta, to be used against biological warfare.

Nerve agent—The Nazis used the first nerve agents: insecticides developed into chemical weapons. Some of the better known nerve agents include VX, sarin, soman, and tabun. These agents are used because only a small quantity is necessary to inflict a substantial damage. Nerve agents can be inhaled or can absorb through intact skin.

Network—A group of components that share information or interact with each other in order to perform a function.

Nodes—A facility at which a chemical is produced, store, or consumed.

Nuclear blast—An explosion of any nuclear material which is accompanied by a pressure wave, intense light and heat, and widespread radioactive fallout which can contaminate the air, water and ground surface for miles around.

Ohm—The unit of measurement of electrical resistance. The resistance of a circuit in which a potential difference of one volt produces a current of one ampere.

Opportunity contaminant—A contaminant that might be readily available in a particular area, even though they may not be highly toxic or infectious or easily dispersed and stable in treated drinking water.

Osama bin Laden (also spelled "Usama")—A native of Saudi Arabia, was born the 17th of 24 sons of Saudi Arabian builder Mohammed bin Oud bin Laden, a Yemeni immigrant. Early in his career, he helped the mujahedeen fight the Soviet Union by recruiting Arabs and building facilities. He hated the United States because he viewed the United States as having desecrated holy ground in Saudi Arabia with their presence during the first Gulf War. Expelled from

Saudi Arabia in 1991 and from Sudan in 1996, he operated terrorist training camps in Afghanistan. His global network Al-Qaeda was credited with the attacks on the U.S. on September 11, 2001, the attack on the *USS Cole* in 2000, and a number of other terrorist attacks. He was killed by U.S. special forces on May 2, 2011.

Owners/operators—Those entities responsible for day-to-day operation and investment in a particular asset or system.

Pandemic influenza—Defined by the World Health Organization (WHO) as a global outbreak of influenza, characterized by an emergent strain of the virus, little to no immunity among the general population, rapid and sustained person-to-person transmission, and lack of a vaccine. On June 11, 2009, WHO determined that 2009 H1N1 influenza (also known as "swine flu") had reached pandemic status.

Pathogen—Any agent which can cause disease.

Pathways—The sequence of nodes and links by which a chemical is produced, transported, and transformed from its initial source to its ultimate consumer.

Peakload—The greatest amount of power given out or taken in by a machine or power distribution system in a given time.

Penstocks—Pipelines or conduits used to convey water under pressure to turbines or pump units.

Performance measure—Indicator, statistic, or metric used to gauge program performance.

Physical security—The use of barriers and surveillance to protect resources, personnel, and facilities against crime, damage, or unauthorized access.

Plague—The pneumonic plague, which is more likely to be used in connection with terrorism, is naturally carried by rodents and fleas but can be aerosolized and sprayed from crop dusters. A 1970 WHO assessment asserted that, in a worst case scenario, a dissemination of 50 kg in an aerosol over a city of 5 million could result in 150,000 cases of pneumonic plague, 80,000–100,000 of which would require hospitalization, and 36,000 of which would be expected to die.

Political terrorism—Terrorist acts directed at governments and their agents and motivated by political goals (i.e., national liberation).

"Possible"—In the context of the threat evaluation process, a water contamination threat is characterized as "possible" if the circumstances of the threat warning appear to have provided an opportunity for contamination.

Potassium iodide—An FDA-approved nonprescription drug for use as a blocking agent to prevent the thyroid gland from absorbing radioactive iodine.

Power—Mechanical or electrical force or energy. The rate at which work is done by an electric current or mechanical force, generally measured in watts or horsepower.

Preparedness—The activities necessary to build, sustain, and improve readiness capabilities to prevent, protect against, respond to, and recover from natural or manmade incidents. Preparedness is a continuous process involving efforts at all levels of government and between government and the private sector and nongovernmental organizations to identify threats, determine vulnerabilities, and identify required resources to prevent, respond to, and recover from major incidents.

Presumptive results—Results of chemical and/or biological field testing that need to be confirmed by further lab analysis. Typically used in reference to the analysis of pathogens.

Prevention—Actions taken and measures put in place for the continual assessment and readiness of necessary actions to reduce the risk of threats and vulnerabilities, to intervention and stop an occurrence, or to mitigate effects.

Prioritization—In the context of the NIPP, prioritization is the process of using risk assessment results to identify where risk reduction or mitigation efforts are most needed and to subsequently determine which protective actions should be instituted in order to have the greatest effect.

Protection—Actions or measures taken to cover of shield from exposure, injury, or destruction. In the context of the NIPP, protection includes actions to deter the threat, mitigate the vulnerabilities, or minimize the consequences associated with a terrorist attack or other incident. Protection can include a wide range of activities, such as hardening facilities, building resilience and redundancy, incorporating hazard resistance into initial facility design, initiating active or passive countermeasures, installing security systems and redundancy, incorporating hazard resistant into initial facility design, initiating active or passive countermeasures, installing security systems, promoting workforce surety, training and exercises, and implementing cyber security measures, among various others.

Psychopath—A mentally ill or unstable person, especially one having a psychopathic personality (*q.v.*), according to *Webster's*.

Psychopathology—The study of psychological and behavioral dysfunction occurring in mental disorder or in social disorganization, according to *Webster's*.

Psychopathy—A mental disorder, especially an extreme mental disorder marked usually by egocentric and antisocial activity, according to *Webster's*.

Psychotic—Of, relating to, or affected with psychosis, which is a fundamental mental derangement (as schizophrenia) characterized by defective or lost contact with reality, according to *Webster's*.

Pumped-storage hydroelectric plant—A plant that usually generates electric energy during peak-load periods by using water previously pumped into an elevated storage reservoir during off-peak, periods when excess generating capacity is available to do so. When additional generating capacity is needed, the water can be released from the reservoir through a conduit to turbine generators located in a power plant at a lower level.

Rapid field testing—Analysis of water during site characterization uses rapid field water testing technology in an attempt to tentatively identify contaminants or unusual water quality.

Rated capacity—That capacity which a hydro generator can deliver without exceeding mechanical safety factors or a nominal temperature rise. In general this is also the nameplate rating except where turbine power under maximum head is insufficient to deliver the nameplate rating of the generator.

Recovery—The development, coordination, and execution of service- and site-restoration plans for affected communities and the reconstitution of government operations and services through individual, private sector, nongovernmental, and public assistance programs that identify needs and define resources; provide housing and promote restoration; address long-term care and treatment of affected persons; implement additional measures for community restoration; incorporate mitigation measures and techniques, as flexible; evaluate the incident to identify lessons learned and develop initiatives to mitigate the effects of future incidents.

Redundancy—An energy reliability strategy based on the notion that multiple systems provide needed backup if one system fails or cannot meet demand.

Reservoir—An artificial lake into which water flows and is stored for future use.

Resilience/resiliency—The ability to resist, absorb, recover from, or successfully adapt to adversity or a change in conditions. In the context of energy security, resilience is measured into terms of robustness, resourcefulness, and rapid recovery.

Response—Activities that address the short-term, direct effects of an incident, including immediate actions to save lives, protect property, and meet basic humans needs. Response also includes the execution of emergency operations plans and incident mitigation activities designed to limit the loss of life, personal injury, property damage, and other unfavorable outcomes. As indicated by the situation, response activities include applying intelligence and other information to lessen the effects or consequences of an incident; increasing security operations; continuing investigations into the nature and source of the threat; ongoing surveillance and testing processes; immunizations, isolation, or quarantine; and specific law enforcement operations aimed at preempting, interdicting, or disrupting illegal activity, and apprehending actual perpetrators and bringing them to justice.

Retentate—In ultrafiltration, the retentate is the solution that contains the particles that do not pass through the membrane filter. The retentate is also called the concentrate.

Ricin—A stable toxin easily made from the mash that remains after processed castor beans. At one time, it was used as an oral laxative, castor oil; castor oil causes diarrhea, nausea, vomiting, abdominal cramps, internal bleeding, liver and kidney failure, and circulatory failure. There is no antidote.

Risk—A measure of the potential for an unwanted outcome resulting from an incident, event, or occurrence, as determined by its likelihood and the associated consequences. Risk should be assessed from any scenario as a function of consequence, vulnerability, and threat. It is important to think of risk as influenced by the nature and magnitude of a threat (T), the vulnerabilities to that threat (V), and the consequences that could result (C).

$$\text{Risk} = f\,(C,V,T)$$

- **Consequence (C) Analysis**: estimates the potential public health and economic impacts that a successful attack could cause
- **Vulnerability (V) Assessment**: identifies weaknesses in a asset design, implementation, or operations that can be explored by an adversary
- **Threat (T) Analysis**: estimates the likelihood that a particular target, or type of target, will be selected for attack, and is based on the intent and capacity of an adversary.

Salmonella—An infection caused by a gram-negative bacillus, a germ of the Salmonella genus. Infection with this bacteria may involve only intestinal tract or may be spread from the intestines to the bloodstream and then to other sites in the body. Symptoms of salmonella enteritis include diarrhea, nausea, fever, abdominal and fever. Dehydration resulting from the diarrhea can cause death, and the disease could cause meningitis or septicemia. The incubation period is between 8 and 48 hours, while the acute period the illness can hang on for 1 to 2 weeks.
Sarin—A colorless, odorless gas. With a lethal dose of .5 mg (a pinprick-sized droplet), it is 26 times more deadly than cyanide gas. Because the vapor is heavier than air, it hovers close to the ground. Sarin degrades quickly in humid weather, but sarin's life expectancy increases as temperature gets higher, regardless of how humid it is.
SCADA (Supervisory Control and Data Acquisition)—A computer-based control system that remotely controls processes previously controlled manually. SCADA allows an operator using a central computer to supervise (control and monitor) multiple networked computers at remote locations. Each remote computer can control mechanical processes (mixers, pumps, valves, etc.) and collect data from sensors at its remote location.
Sector—A logical collection of assets, systems, or networks that provide a common function to the economy, government, or society. The NIPP addresses 18 Critical Infrastructures sectors, identified by the criteria set forth in HSPD-7.
Sentinel Laboratory—An LRN lab that reports unusual results that might indicate a possible outbreak, and refers specimens that may contain select biological agents in reference labs within the LRN.
Site characterization—The process of collecting information from an investigation site in order to support the evaluation of a drinking water contamination threat. Site characterization activities include the site investigation, field safety screening, rapid field testing of the water, and sample collection.
Situational awareness—An understanding of the current environment and the ability to accurately anticipate future problems in order to respond effectively.
Sleeper cell—A small cell which keeps itself undetected until such time as they can "awaken" and cause havoc.
Smallpox—The first biological weapon, used during the eighteenth century, smallpox killed 300 million people in the nineteenth century. There is no

specific treatment for smallpox disease, and the only prevention is vaccination. This currently poses a problem, since the vaccine was discontinued in 1970 and the WHO declared smallpox eradicated. Incubation is 7 to 17 days, during which the carrier is not contagious. Thirty percent of people exposed are infected, and it has a 30 percent mortality rate.

Smart grid—The electric delivery network, from electrical generation to end-use customer, integrated with the latest advances in digital and information technology to improve customer, integrate with the latest advances in digital and information technology to improve electric-system reliability, security, and efficiency.

Sociopath—Basically synonymous with psychopath (*q.v.*). Sociopathic symptoms in the adult sociopath include an inability to tolerate delay or frustration, a lack of guild feelings, a relative lack of anxiety, a lack of compassion for others, a hypersensitivity to personal ills, and a lack of responsibility. Many authors prefer the term sociopath because this type of person had defective socialization and a deficient childhood.

Sociopathic—Of, relating to, or characterized by asocial or antisocial behavior or a psychopathic (*q.v.*) personality, according to *Webster's*.

Spore—An asexual, usually single-celled reproductive body of plants such as fungi, mosses or ferns; a microorganism, as a bacterium, in a resting or dormant state.

Supply chain—In the Energy Sector, a system of interdependent processes and assets that includes the extraction of raw material, the refinement and transportation of fuel, the generation, transmission, and distribution of electricity, and the delivery of energy, in its various forms, to the end-use consumer.

System—Any combination of facilities, equipment, personnel, procedures, and communications integrated for a specific purpose.

Target analysis—Estimates the likelihood that a particular target, or type of target, will be selected for attack, and is based on intent and capability of an adversary.

Terrorism—Premeditated threat or act of violence against noncombatant persons, property, and environmental or economic targets to induce fear, intimate, coerce, or affect a government, the civilian population, or any segment thereof, in furtherance of political, social, ideological, or religious objectives.

Terrorist group—A group which practices or has significant elements which are involved in terrorism.

Threat—An indication that a harmful incident, such as contamination of the drinking water supply, may have occurred. The threat may be direct, such as a verbal or written threat, or circumstantial, such as a security breach or unusual water quality.

Threat capability—The availability or the ease of use of tools or methods that could potentially be used to damage, disrupt, or destroy critical functions.

Toxin—A poisonous substance produced by living organisms capable of causing disease when introduced into the body tissues.

Transponder—A device on an airliner which sends out signal allowing air traffic controllers to track an airplane. Transponders were disabled in some of the planes high jacked on 9/11.

Transportation Security Administration—A new agency created by the Patriot Act of 2001 for the purpose of overseeing technology and security in American airports.

Tularemia—An infectious disease caused by a hardy bacterium *Francisella tularensis,* found in animals, particularly especially rabbits, hares, and rodents. Symptoms depend upon how the person was exposed to tularemia but can include difficulty breathing, chest pain, bloody sputum, swollen and painful lymph glands, ulcers on the mouth or skin, swollen and painful eyes, and sore throat. Symptoms usually appear form 3 to 5 days after exposures but sometimes will take up to 2 weeks. Tularemia is not spread from person to person, so people who have it need not be isolated.

Turbine—A machine for generating rotary mechanical power from the energy of a stream of fluid (such as water, steam, or hot gas). Turbines convert the kinetic energy of fluids to mechanical energy through the principles of impulse and reaction, or a mixture of the two.

Ultrafiltration—A filtration process for water that uses membranes to preferentially separate very small particles that are larger than the membrane's molecular weight cut-off, typically greater than 10,000 Daltons. (A Dalton is a unit of mass, defined as 1⁄12 the mass of a carbon-12 nucleus. It's also called the atomic mass unit, abbreviated as either "amu" or "u").

Value proposition—A statement that outlines the national and homeland security interest in protecting the nation's critical infrastructure and articulates the benefits gain by all critical infrastructure partners through the risk management framework and public-private partnership described in the NIPP.

Vector—An organism which carries germs from one host to another.

Vesicle—A blister filled with fluid.
Volt (V)—The unit of electromotor force or potential difference that will cause a current of 1 ampere to flow through a conductor with a resistance of 1 ohm.
Vulnerability—A physical feature or operational attribute that renders an entity open to exploitations or susceptible to a given hazard.
Watt (W)—The unit used to measure production/usage rate of all types of energy; the unit for power. The rate of energy transfer equivalent to one ampere flowing under a pressure of one volt at unity power factor.
Watt-hour (Wh)—The unit of energy equal to the work done by 1 watt in 1 hour.
Weapons of mass destruction—According to the National Defense Authorization Act, any weapon or device that is intended, or has the capability, to cause death or serious bodily injury to a significant number of people through the release, dissemination, or impact of

- toxic or poisonous chemicals or their precursors
- a disease organism
- radiation or radioactivity

Xenophobia—Irrational fear of strangers or those who are different from oneself.
Zyklon b—A form of hydrogen cyanide. Symptoms of inhalation include increased respiratory rate, restlessness, headache, and giddiness followed later by convulsions, vomiting, respiratory failure and unconsciousness. Used in the Nazi gas chambers in World War II.

Index

www.ingramcontent.com/pod-product-compliance
Lightning Source LLC
LaVergne TN
LVHW090316210926
848480LV00001B/264
9781598888737